Fired Up!

The Optimal Performance Guide for Wildland Firefighters

Fired Up!

The Optimal Performance Guide for Wildland Firefighters

Charles Palmer, Ed.D.

Birch Grove Publishing

Copyright © 2007 Birch Grove Publishing. All rights reserved. Except as permitted under the United States Copyright Act of 1976, no part of this publication may be reproduced in any form or by any means, or stored in a data base or retrieval system, without the prior written permission of the publisher. Write to: Permissions, Birch Grove Publishing, PO Box 131327, Roseville, MN 55113.

Publisher: Paul Nockleby
Cover Design: Dawn Mathers Design
Cover Photo: Robin Loznak/The Daily Inter Lake

Chapter 4 was originally published in *Under Fire: The West is Burning* by Fenske Media. Reprinted by permission.

Library of Congress Cataloging-in-Publication Data

Palmer, Charles
    Charles Palmer
    232 pp. 15 cm.
    Includes bibliographical references and index.
    ISBN 978-0-9744071-4-2
    1.   I. Title
            2007

Manufactured in the U.S.A.
1 2 3 4 5 6 7 8 9 10 11 12 13 14 15 16

Contents

	Acknowledgments	6
1	Introduction	7
2	Exercise Physiology and Sport Psychology	10
	PART ONE: PHYSICAL FACTORS IN OPTIMAL PERFORMANCE	18
3	Physical Training	20
4	Nutrition, Hydration, and Ergogenic Aids	79
5	Injury	108
	PART TWO: PSYCHOLOGICAL FACTORS IN OPTIMAL PERFORMANCE	125
6	Attitude	127
7	Goal Setting	136
8	Teams	144
9	Stress	167
10	Transitions	179
	PART THREE: ENVIRONMENTAL FACTORS IN OPTIMAL PERFORMANCE	190
11	Relationships	193
12	Pulling It All Together	221
	Book References	227
	Index	230

Acknowledgments

In preparing this book, the author gratefully acknowledges the assistance offered from the following sources: Don Haugo for helping to connect publisher and writer, Tommy Carr for his expertise in helping to develop the Firefighter Performance Workout and for appearing in the workout photos, Sara Shermann for appearing in the workout photos, Dr. Steve Gaskill for general consulations, the professionals at Mission-Centered Solutions, Lee Nelson of the U. S. Fish and Wildlife Service in Minnesota, and all of the firefighters who took the time to talk with me and fill out the relationship surveys.

The author also wishes to thank wildland firefighters everywhere for doing what they do, with special props going out to the Missoula Smokejumpers and the rookie class of 1995. This guide is for firefighters worldwide, and the author hopes that it can be of assistance. And finally, without the encouragement, support, and patience of my wife Christine, this book could not have been completed.

1
Introduction

During a typical year, an estimated 100,000 to 200,000 wildland fires burn within the United States. These fires blacken thousands of square miles, and cost hundreds of millions of dollars to extinguish. In the record-breaking 2000 fire season, 122,827 fires burned more than 8.4 million acres (figures from the National Interagency Fire Center, www.nifc.gov). Wildland fires are no respecter of boundaries, either. Canada, Australia, the former Republics of the Soviet Union, France, and Portugal are just a few of the countries around the world which also contend with significant wildfires each and every year.

In the U.S., thousands of firefighters battle wildland blazes each year, responding from federal, state, and local governmental agencies as well as private contractors and volunteer departments. Our job as firefighters: to provide for our own safety while simultaneously protecting the public and the natural resources of the country we are in. This is a tall order considering how intense and active fire seasons have become in the past few years.

Fighting wildland fires demands tremendous strength and stamina, and so most wildland firefighters spend considerable time and energy getting into shape, physically. Many of us are even paid to engage in physical training (PT) during the workday. We run, lift weights, bike, or hike — all in an effort to prepare ourselves for the rigors of the fireline. There's good reason for all the training. If we are faced with retreating from a blowup, for example, physical conditioning can spell the difference between life and death.

On fires, we could be required to dig a fire line, chain after chain, over country that would leave a mountain goat gasping for breath and looking for a shortcut. Some firefighters operate chainsaws for hours on end, or drag pumps and hose through hell and high water, or carry heavy smokechaser packs, looking for that elusive fire spotted by a distant lookout. Shifts are long, the work is dirty and difficult, and at the end of the day what often awaits us is a meal of questionable quality and a thin sleeping bag rolled out in the dirt. Our bodies have to be prepared for the physical rigors and deprivations of the fireline.

That being said, there is much more to firefighting than just **physical performance**. In my opinion, most firefighters need to better address the **psychological factors** and **social factors** associated with fighting fires. Although most of us are good at physical training (PT), many of us neglect mental training (MT) for our minds. By mental training, I mean such things as group dynamics, teamwork, cohesion, relationships with co-workers and significant others, dealing with interpersonal and intrapersonal stress, dealing with injuries, time management and goal-setting, and career transitions, to name a few. Ironically, when things go bad on firelines, the cause is most usually traceable to human (mental) errors, not physical ones. Therefore, in order to become more effective firefighters, we need to address all of the performance components that are involved with wildland fire suppression.

Fortunately, there are bodies of knowledge about firefighting that can be brought forward to help us. On the physical side, research and information from the field of exercise physiology can help us understand how the human body functions, thus assisting in the development of training programs to meet the unique physical challenges of our profession. Nutritional and hydration needs for wildland firefighters also need to be covered, since physical performance involves much more than just a person's level of conditioning.

With regard to the mental aspects of the job, information from sport psychology, also called performance psychology, can be applied to wildland firefighters. Although sport psychology's primary focus has been on athletes and those who coach them, much of what has been developed for them over the past two decades applies just as well to firefighters. The similarities that firefighters share with elite athletes far outnumber the differences. This concept will be expanded in chapters that follow.

In sum, the overall purpose of this guide is to look at firefighter performance from a holistic perspective, focusing not only on the physical side of the equation, but at the psychological and environmental components as well. Utilizing information and concepts from the fields of performance psychology and exercise physiology, a unique model of optimal performance is presented. It is hoped that the reader will gain a better understanding of the entire range of skills and knowledge involved in performing at an optimal level, and will see ways to relate this material to the world of wildland firefighting.

So, looking ahead, a brief review of the fields of exercise physiology and sport psychology in Chapter Two prepares the reader and leads up to the heart of the book, which are the physical, psychological, and environmental factors of performance, discussed in Chapters Three through Eleven. Exercise physiology and sport psychology are necessary brush-clearing for what is incorporated within these core topics at the heart of the book, a topic we return to once again in the summary discussion of Chapter Twelve.

Exercise Physiology and Sport Psychology

Exercise Physiology (ExPhys) is the in-depth study of physiological responses of humans to exercise and physical conditioning. It is an extremely useful field for wildland firefighters, because exercise physiologists (EPs) do all of the following:

- look at human energy outputs during periods of exertion and compare them to times of rest

- focus on energy and nutrition, and consider the role these have in athletic performance as well as overall health

- investigate the body's systems during physical exertion, such as the pulmonary, cardiovascular, neuromuscular, and endocrine systems

- explore training programs for muscular strength, and develop ways to train the muscles for both anaerobic (without oxygen) and aerobic (with oxygen) power

- look into the impact of environmental factors such as heat, cold, and altitude

- analyze the role of ergogenic aids, such as caffeine, nicotine, bicarbonate drinks, and anabolic steroids, to see what role they play in human performance

Exercise Physiology has much to offer wildland firefighters, with its emphasis on physical conditioning, nutrition, and energy expenditure. These are the "upside" of physiology – the body at its best – and Chapters Three and Four focus on them. But the "downside" of physiology is injury, not to mention aging, disability, and death. Firefighting is intrinsically hazardous work, and injury is a perpetual occupational hazard. Chapter Five provides vital information on prevention and treatment for injuries should they occur. All of these topics are gathered into the first section of the book under the rubric we call **Physical Factors**.

What is Sport Psychology?

The field of psychology boomed after 1945 following the end of World War II. Besides the physical casualties of war in millions of dead and wounded, the war exacted an emotional toll on millions of other veterans. Psychologists were recruited to help heal the psychological wounds of war among those directly involved (soldiers, nurses, and civilians near the frontline).

Subspecialties also emerged during this boom in the field. For example, clinical psychology, industrial-organizational psychology, forensic psychology, and school psychology began to form as specialties within the overall field. As noted by Murphy (1995), these subspecialties were defined primarily by setting and type of clients. For example, school psychology evolved to work in schools with students; and industrial-organizational psychology developed as a subspecialty focusing on employers and employees in companies and organizations.

Before 1960, very few psychologists worked exclusively with athletes. The small number who did seemed to do it with little specialized training in sport. By the late 1960's and 1970's, however, a small but increasing number of psychologists began focusing on athletes. The reasons for this growth are recounted by Murphy (1995):

> A major factor in the increasing involvement of psychologists with sport organizations has been the increasing professionalization of sport itself. The advent of two factors in the 1960's and 1970's transformed the face of sport in America: sports television and sports sponsorship. Money was the common element. As networks paid organizations to air their competitions and as companies paid them for the right to associate products with star athletes, the profitability of sport organizations grew. Athletes in turn began to receive substantial sums for their sports participation. This trend was evident across all areas of major sport: professional, collegiate, and Olympic. Although many lament the demise of "playing for the love of the game," the influx of money into the sporting arena was inevitable. Sport organizations began to view athletes as valuable assets. Without great athletes, success was out of reach — success that bred bigger television contracts and bigger sponsorship deals. An organization that had spent years developing a scouting and coaching system to create successful athletes was averse to losing them, especially to factors perceived as psychological problems. Thus psychologists were hired as consultants to work with athletes, in a sense protecting the investment of the organization. Although some might decry this analysis as overly economic, ample evidence suggests it is a viable model that explains the development of sport over the last twenty years. (p. 3)

By the 1980's, sport psychology had grown even more, as researchers sought new and better ways to apply the techniques of psychology to athlete populations. In other words, efforts were underway to "field test" some of these new "tools." Psychologists began asking themselves: How can we take these theories and apply them so that they can be of benefit to athletes? In the years since 1990, sport psychology has developed into one of the fastest growing subdivisions within the field of psychology.

What Do Exercise Physiology and Sport Psychology Have To Do With Firefighters?

We said in the introduction that firefighters have much in common with athletes. The similarities are not just with weekend athletes, but with elite and professional athletes as well. Consider some of the traits and lifestyles of top athletes:

- Their livelihood depends on peak physical conditioning
- They have months of intense work followed by months of layoff
- They are paid to perform extremely difficult physical tasks
- They are carefully watched by the news media and the general public
- Their risk of injury on the job is far higher than other occupations
- They travel around the country, usually within teams, deploying their equipment and skills in a variety of different arenas, with different weather conditions, and different times of day and night
- At some point (called "retirement", when they no longer have the desire or ability to compete at a peak level, but long before they are old enough to collect Social Security) they must transition out of a career they may love and begin a new phase of life

There are obviously many points of similarity between athletes and wildland firefighters. Granted, firefighters do not get paid the wads of money that professional athletes get paid for each minute of performance. However, the traits and lifestyles of professional athletes and wildland firefighters do have many other points in common. Above all, our livelihood and even our lives as firefighters depend on a high degree of physical conditioning. We are paid to perform very rigorous and demanding physical tasks that average Joes cannot perform: hiking long distances in a hurry in rocky terrain, digging fire lines with hand tools, hauling heavy equipment in heavy or soggy boots for miles, and so on.

At the same time, our job becomes more high-profile each year in the media, especially with the growing intensity of recent fire seasons and the assignment to branch out into different "all-risk" missions. We have a "season" just like athletes do. Many of us travel around the country (often times in planes and buses) from one fire to the next. Most often we work in teams — be they small teams (such as a helitack crew or an engine module) or large teams (20-person Type I or Type II crews, and Incident Management Teams). Lastly, just as athletes retire from their sport, we must leave the fireline to others at some point, be it retirement or when we take a job outside of fire suppression.

Again, the most obvious similarity between firefighters and athletes are the physical demands of the job. Therefore, it only makes sense that the field of performance psychology which helps athletes do their best should also offer useful insights to firefighters. One of the key tenets of performance psychology consists in helping individuals and teams attain optimal performance, or also called peak performance. That is, in whatever arena in which people operate, how can they perform to the best of their ability, as consistently as possible? An introduction to what is called "the peak performance model" helps us grasp this key concept.

The Peak Performance Model (PPM)

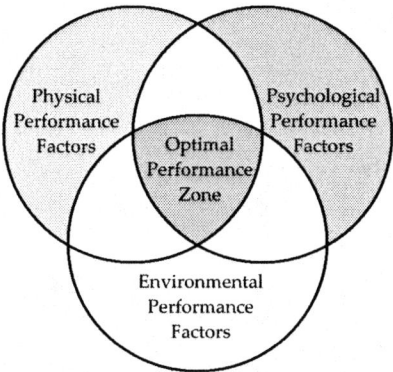

According to the Peak Performance Model, human functioning is divisible into three distinct categories:

- physical performance factors — such things as our level of fitness, as well as physical abilities such as strength, agility, and coordination

- psychological performance factors — such traits as our personality type, motivation, mental toughness, and internal drive

- environmental performance factors — family, friends, significant others, teachers, bosses, teammates, rules, policies, and support systems — all those things which surround us in our daily lives

On the Peak Performance Model, the area where physical, psychological, and environmental factors intersect is considered to be the zone of optimal performance. Notice on the diagram that the zone of optimal performance in the middle part of the diagram is a relatively small area when compared with the totality. In order for the firefighter to achieve optimal or peak performance, all three areas must be at their peak.

Looking at the Peak Performance Model sector by sector:

- **physically**, we must be at the top of our game — our bodies must be able to handle the demands of wildland fire suppression: long hours, arduous work, little food, and poor sleeping conditions

- **psychologically**, we need to be alert and prepared for whatever challenge comes our way, whenever and wherever and whatever the source

- **environmentally**, we need to be getting all the support and guidance that is necessary for us to function as effectively as possible

It might to help to compare this Peak Performance Model to the Fire Triangle. Remember, the fire triangle consists of heat, fuel, and oxygen. All three are needed for fire to occur. Removing just one of these elements eliminates the possibility of combustion. The Peak Performance Model functions roughly the same way. All three factors — physical, psychological, and environmental — are needed for optimal performance to occur. Take away or diminish just one of the three factors, and we are not performing at our optimal level.

A more suitable model of optimal performance for firefighters might look like this:

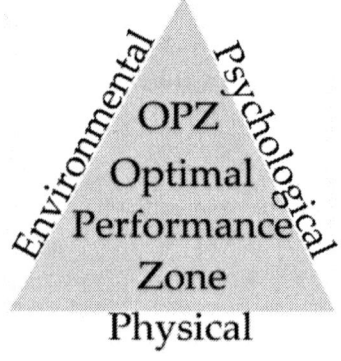

In the world of fire suppression, "OPS" is short for "Operations", an integral part of any Incident Management Team. For Peak Performance, let the acronym **OPZ** stand for the **Optimal Performance Zone**, which occurs when we are hitting on all cylinders of physical, psychological, and environmental performance factors.

Working in a high-risk environment like wildland fire, we oftentimes need to be performing at or very near our peak level. When we are not performing at that level, negative consequences can occur. Something below peak performance may simply result in the fire not being stopped as quickly as possible, or the team not performing up to its potential. However, significantly poor performance can lead to injury or death for you or the people who work with you. An athlete who is not functioning optimally can shake off a poor performance and get ready for the next competition. But the firefighter with a significantly poor performance on the road may not be able to "coast" to the next fire or incident.

As mentioned before, firefighters do a good job of attending to the physical factors that are associated with our job. We physically train regularly and, for the most part, take pretty good care of our bodies (except for drinking binges when we finally get a day off after 14 days on a fire).

So, while most of us generally take good care of our bodies, many of us need to pay greater attention to the psychological and environmental performance factors in our daily routines. We should remember the OPZ and strive to reach it each and every day we go to work. It is hoped that this guide will help in your pursuit of optimal performance on the fireline.

Part One: Physical Factors in Optimal Performance

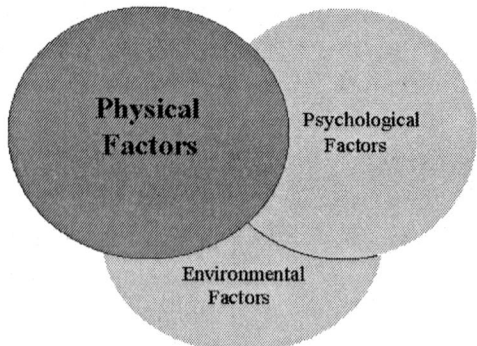

Most every firefighter recognizes the importance that physical conditioning plays with regard to our performance on the fireline. After all, wildland fire suppression is often an extremely difficult job in demands placed upon our bodies. As noted in Chapter Two, the life of a wildland firefighter is in many ways similar to that of an elite athlete, in that we are paid to perform physical tasks. To prepare for this, most firefighters engage in some sort of physical training (PT) program since they realize that shifts will be long, packs heavy, and the miles to hike at times may be many. It is no accident that in the Optimal Performance Triangle, physical factors form the base leg of the model.

Like the roots and trunk of the tree, physical factors are the foundation upon which everything else rests.

With such importance placed upon physical conditioning, one might assume that comprehensive, standardized training programs would be in place to help firefighters prepare for the types of working conditions that we encounter. Unfortunately, that is not the case. Governmental agencies, be they national, state, or local, have provided little or no guidance to their wildland firefighting forces on how best to train for wildland fire suppression. Consequently, physical training programs must often be developed by each district or individual unit, or by the employees themselves. Some districts, units, and employees do an excellent job of developing their own physical training regimens. But that is not the case across the board.

With scant support or guidance available, physical training programs can become disjointed or irrelevant, with activities that have little relevance to what we actually do on the fireline. Sessions of physical training are often left to each individual's discretion. And so, some firefighters run, some lift weights, others bike or walk on a treadmill, while a small percentage choose to do as little as possible. As firefighters, we must begin to ask ourselves the question: How effectively are we preparing our bodies, physically, for what we actually have to do on the job? If the answer to this question is "little", "nothing" or "not sure," then maybe the information in the following section will be of help.

Since physical performance depends on a whole lot more than just our overall level of conditioning, this section will look not only at physical training, but also at nutrition, hydration, the use of ergogenic aids, and to the prevention and recovery from injury. Our body is the most important tool we have as firefighters. The sharper and stronger that tool is, the better able we will be to accomplish our tasks and missions.

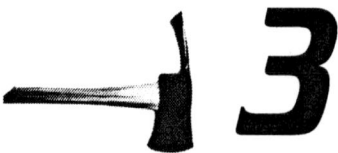

Physical Training

April 1995: The day broke clear and cold as Jeff and I navigated the winding road around Holter Lake, the body of water formed in this central part of Montana when the Missouri River was dammed in the 1910's. Our destination, Mann Gulch, lay ahead of us. In August of 1949, twelve Missoula smokejumpers and a recreation guard died there when a routine fire suddenly blew up and caught them in the tall grass that had been cured by record high temperatures.

Reaching the end of the road, we parked the truck and began sorting through our packs. By choice, my load would be heavy, consisting of a couple plastic canteens of water, two sandwiches, a few other items, and fifty pounds of weight plates I had borrowed from the local gym. Since reading the book "Young Men and Fire," I had wanted to see the place with my own eyes, to view firsthand the ground where these firefighters I looked up to had perished. In less than two months, I would begin rookie smokejumper training at the same base from which these men came. If I was going to make it through the program, I needed to be in much better shape than I was currently in. Thus, the extra weights in the sixty-pound pack that hung heavily from my shoulders.

After putting on our fishing waders, we crossed Willow Creek. We then stashed the waders in the brush on the other side of the creek. The hike in to Mann Gulch was tough, made more difficult by the weight I was carrying. Two and a half miles later, we crested the top of Rescue Gulch. From our vantage point, we quickly spotted the lone white cross on the other side of the draw. This simple memorial marked the spot in Rescue Gulch where Bill Hellman had fallen after being burned in Mann Gulch, located just to the south. The next day, tragically, Hellman would die from his burns in a Helena hospital.

Reaching Hellman's cross, I dug down into my pack and pulled out one of the thirteen roses I had purchased at a Great Falls shop. I placed it by the marker, and murmured an awkward word of thanks for his sacrifice. After descending into Mann Gulch, Jeff and I repeated the ritual at each of the twelve other markers that we found. By the time we finished, the muscles in my legs were throbbing. No amount of pain I felt, though, could compare with the misery these men had faced back in 1949.

We sat down to eat our lunches, and afterward started the long hike back out. A little more than an hour later, we reached the waders we'd stashed, and we then crossed the creek once again. Finally getting back to the truck, Jeff and I dumped our packs into the back end, and slowly climbed into the cab. For what seemed an hour on the drive out, neither of us spoke, each physically drained and emotionally spent by our day in the Gulch. I still had a long way to go to get my body into shape if I was going to follow in these men's footsteps.

It is frequently stated in this guide that wildland firefighting is a physically challenging occupation. The National Wildfire Coordinating Group (NWCG), the agency responsible in the United States for standards of training, equipment, qualifications, and other operational functions in wildland fire, refers to it as **arduous work**. In the 310.1 Wildland and Prescribed Fire Qualification System Guide, the NWCC states:

> *Duties involve field work requiring physical performance calling for above-average endurance and superior conditioning. These duties may include an occasional demand for extraordinarily strenuous activities in emergencies under adverse environmental conditions and over extended periods of time. Requirements include running,*

> *walking, climbing, jumping, twisting, bending, and lifting more than 50 pounds; the pace of work typically is set by the emergency condition.*

Obviously, in order to meet these demands, firefighters need to be in good physical shape. From the off season, to getting prepared to take the Work Capacity Test (WCT) prior to getting a red card for the year (if you are an American firefighter), all the way through the busy months of the fire season, staying in good physical shape is a year-round objective for most of us in the profession. The better condition we are in physically, the more ready we are to meet the arduous work demands and the more likely we are to avoid injury. Fortunately, most firefighters are active people, which means they maintain a good overall level of fitness. However, judging from personal experience and observation, many firefighters do not train as effectively and efficiently as they could. This guide hopes to address that deficiency.

There is an old military saying which simply states: Train like you fight. In other words, training regimens need to simulate as closely as possible the activities we engage in during regular operations. Put another way, our training needs to match the fire environment. So, for example, how closely does a certain firefighter's workout (light stretching, bench presses, sit-ups, and a three mile run) match the heavy demands that are placed on his or her body when attacking a two-acre fire? Granted, the person has engaged in physical training, which helps improve their overall level of conditioning. But how often on the fireline are you asked to lie on your back and press one hundred pounds? When was the last time you had to run long distance on a fire?

So the first step is to relate the training you need to do to your own experiences on the fire ground. From a physical standpoint, firefighting is most often about hiking long distances, in difficult terrain, while carrying a heavy pack. It is about bending over for long periods of time while digging in the dirt, or running a chain saw for hours on end. Fighting fire involves lots of lifting (cubies, trees, pumps, various equipment), and a great many activities which put your body into semi-awkward positions. Fighting fires at times can entail short bursts of feverish activity (intense hotline construction or a tough hill climb) followed by longer periods of lower intensity work (mop-up).

This is not meant to imply that running or bench presses have no place in the workouts we do. Each of these can be viable components of a physical training regime. Rather, the point is that if they are done, these training components must be seen as small parts of an overall program that focuses on training our bodies in ways that mimic what we do on the job. More details on how this is done later. For now, as most readers know, typical workout programs consist of two main parts: strength training and cardiovascular training. While these are arguably the most important, effective physical training for firefighters must also involve other vital components so that the body is best prepared to engage in exercise and recover from exercise once it has been completed. Together, they form what I call the Firefighter Performance Workout (FPW).

Firefighter Performance Workout

Most firefighters are allowed to exercise as part of their daily job duties. However, this time is usually limited. Firefighters also tend to lead rather busy lives, so when working out during the off-season or on weekends, many would probably prefer to not spend two or three hours in the gym trying to get into shape. With this in mind, the FPW has been designed to encompass a complete full-body workout within a 60-75 minute time period. Also, since firefighters tend to be rather individualistic and independent in nature, the FPW simply provides a number of different activities and suggestions, and allows each individual to tailor the workout to his or her own preferences and needs.

The FPW consists of five phases: **Warmup, Flexibility, Cardiovascular Training, Core/Strength Training**, and **Cool Down**. Each of these are elaborated in detail in this book. A word of caution: every reader should consult with his or her physician or medical caregiver before beginning this or any other physical training program. Enough with the legal throat clearing; let's get started!

Firefighter Performance Workout (FPW)

Warmup: 5-7 minutes
Flexibility: 10 minutes
Cardiovascular: 20-30 minutes
Core/Strength Training: 20-30 minutes
Cool Down: 5-10 minutes

Warmup (5-7 minutes)

In many ways, your body is like an older-model car (for some of us, a really old used car!). Now, when you start an older car, you usually let them idle for awhile (warmup) before you put them in gear and start pushing the performance envelope. Just as its name implies, the physical warmup is simply an attempt to get the body prepared for the workout session it is about to undertake. The goal in the warmup phase is to involve as much of the body as possible, so that most, if not all of the major systems are being exercised. Jogging or slow running will do, as will biking. Better yet will be the elliptical-type machines, cross-country ski trainers (such as NordicTrak™), or rowing machines, since they all require more involved upper body motions in addition to the leg workout.

The warmup should be done at a low to low/moderate intensity, resulting in a light sweat and an increased heart rate. If done effectively, five to seven minutes should be adequate to prepare the body for PT.

Flexibility (10 minutes, 3 times per week)

Let's face it. Very few people stretch adequately before exercising. With the possible exception of those who are rehabilitating some sort of injury, most of us simply launch into our workouts without first engaging in any kind of stretching exercises. Yet physiologists and exercise professionals have known for decades that flexibility is an important component of any training program, since flexible muscles performs better than tight ones and are much less prone to injury.

Simply defined, flexibility is the ability to move a joint through its complete and natural range of motion. Flexibility exercises promote increased muscle length, improved joint lubrication, and improved elasticity of connective tissues. Injuries can occur when a limb or joint is forced beyond its normal range of motion. Therefore, by improving range of motion in your limbs and joints, flexibility training can help to reduce the potential for injury.

A great number of stretches can be done in a short amount of time, if they are done efficiently. The objective of the FPW is a complete body workout, and so the goal in warming up is to stretch as many different parts of the body as possible. The firefighter should look through the following stretches and select those which achieve this goal. Also, since flexibility training is recommended three times per week, firefighters working out on a daily basis can devote the extra ten minutes to other areas of the FPW, or simply shorten their workout by ten minutes on that day.

Some rules to remember when doing flexibility training:

- Perform stretching exercises when your body is warm. This is one of the reasons why **warmup** is the first component of the FPW. Five to seven minutes of light aerobic exercise will help the body prepare for the stretching of muscles it is about to undertake.

- Try to stretch as many muscle groups as possible. Focus particularly on stretching those muscles that are most often used in our line of work, for example, legs, back, trunk, shoulders.

- Hold each stretch for 20-30 seconds or longer for those parts of the body that are more tight. Ease slowly in and out of each stretch, and **do not bounce!**

- Repeat the same stretch 2-3 times for any areas of the body that are particularly tight.

- Try to get in the habit of stretching out prior to engaging in any physically demanding activity, not just before working out. This might include such activities as a strenuous hike into a fire, chainsaw work, or heavy lifting. A few minutes stretching may save days or weeks of recovery from an injury suffered because your body was not ready for exertion.

- This should be a no-brainer, but **stop** stretching if you feel any pain whatsoever!

The number of stretches is nearly limitless. On the following three pages are eight stretches to warm up the major muscle groups of the body.

Physical Training / 27

Chest Stretch

Calf Stretch

Groin Stretch

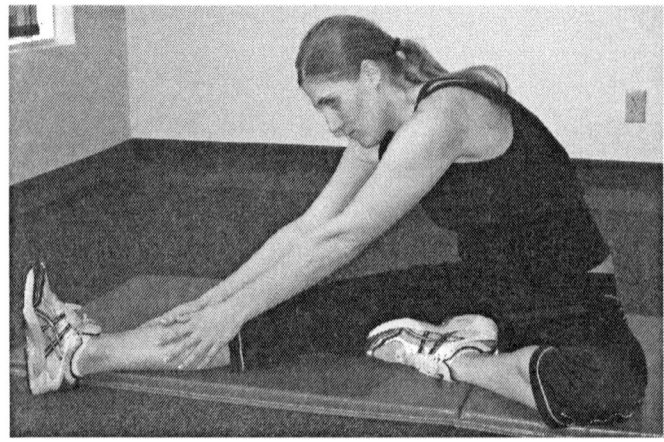

Leg Stretch (each leg)

Back/Hamstring Stretch

These stretches show ways to quickly and efficiently warm up the major muscles groups of the body. Feel free to incorporate your own stretches if you have particular favorites.

Physical Training / 29

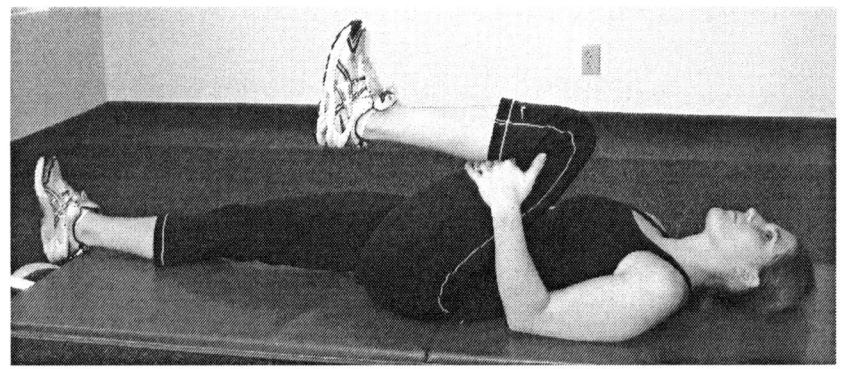

Hamstring Stretch

 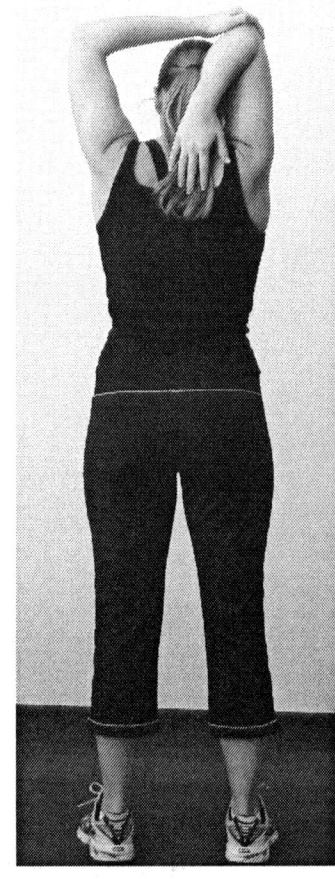

Shoulder Stretch   Triceps Stretch

Cardiovascular Training (20-30 minutes)

Cardiovascular fitness is simply defined as the body's ability to get oxygen and blood to the muscles. Any physical activity that results in an elevated heart rate and deeper breathing is considered to be cardiovascular in nature. Also referred to as aerobic exercise, cardiovascular training has a host of positive benefits: gains in muscle size and strength, loss of fat, increased energy, increases in HDL (good) cholesterol, lower blood pressure, improved immune system functioning, and protection against a wide variety of diseases, including heart disease, stroke, hypertension, diabetes, and osteoporosis (condition in which bones become fragile and more likely to break). Psychological benefits include a greater sense of well-being and decreases in anxiety and depression.

No matter what form of cardiovascular training is done, it is generally categorized according to two principles: duration (how long) and intensity (how hard). Most experts agree that cardiovascular training, to be of benefit, should be done around 3-4 times per week for periods of 15-50 minutes, with the amount of training correlated to the intensity. Intensity is most often defined as a certain percentage of the maximum heart rate (HR).

Maximum heart rate is simply the theoretical number of beats per minute that your heart is capable of producing. It is found by subtracting your age from the number 220 (the generally accepted figure for the human body's maximum heart beats per minute capability, although some individual's may have a higher number than this). Therefore, for a thirty-year-old individual, that person's maximum heart rate would be 220 minus 30 = 190 beats per minute. Remember, this is a rule of thumb, and not a prescription. Again, be sure to consult your doctor before beginning any regimen, and check this figure against what he or she recommends in your particular case.

As mentioned above, intensity is usually defined as a percentage of this maximum heart rate. A person's target heart rate is the range of heart beats per minute at which she or he should work at in order to best achieve aerobic fitness. Typically, this range is between 60% and 85% of a person's heart-rate maximum. Training at 60% of maximum heart rate is considered "low intensity", while workouts at the 80% to 85% level are considered "high intensity". Many pieces of cardiovascular training equipment have built in heart-rate sensors, so it is quite simple to see where your heart rate is at when working out. Heart rate monitors can also be beneficial when attempting to determine heart rate. A low-tech measure is to simply take your pulse for ten seconds during a workout (stop working out if you can't get a reading). Multiply this ten-second pulse reading by six for your heart rate.

For those who are not really interested in what their heart rate is during workouts, the Borg Rating of Perceived Exertion might be an option for measuring the intensity of physical activities. "Perceived exertion" is a subjective measure of how hard you feel your body is working. The Rating of Perceived Exertion (RPE) is based on the physical sensations a person experiences during cardiovascular activity, including such things as increased heart rate, increased respirations or breathing rate, increased sweating, and muscle fatigue. Through self-monitoring of how hard your body is working, you can adjust the intensity of the workout by speeding up or slowing down your movements.

Although there are several different Borg scales, the most commonly used version operates on a continuum from 6 to 20, where 6 is no exertion at all and 20 is maximal effort. The midpoint is 13, which is quite a workout, but one feels one can continue. A rating of 9 is very light work, while at the rating of 17 healthy people feel they can go on, but really have to push themselves. A rating of 19 corresponds to the most strenuous exercise people have ever experienced.

The complete Borg scale is as follows:

Borg Rating of Perceived Exertion

5: No exertion at all
6:
7: Extremely light exertion
8:
9: Very light - easy walking
10:
11: Light workout
12:
13: Somewhat hard, moderate workout
14:
15: Hard workout
16:
17: Very hard, strenuous, tiring workout
18:
19: Extremely hard; cannot sustain for long
20: Maximal exertion

By monitoring how your body feels during workouts, it will become easier to identify your level of exertion. When appraising this, be honest with yourself, regardless of what the actual physical load might be. The RPE is a personal measure. A 12 for you might be somebody else's 17. Exercise professionals generally agree that perceived exertion rates of 12-14 suggest that the workout is being done at a moderate level of intensity. Most firefighter workouts should probably be done at or near this level. However, workouts periodically need to be at a higher intensity, since the fires we suppress oftentimes demand more exertion. By following this pattern, we are training like we fight.

The question still remains: What is the best form of cardiovascular training for wildland firefighters? With so many options available, this might seem to be a difficult question. However, if we remember the maxim that we train to meet the challenge, the most effective form of cardiovascular exercise that firefighters can do is hiking in uneven terrain carrying a weighted pack. Since this is usually what we are called upon to do on the job, this kind of training best matches what we will probably have to do in the field (plus it is great preparation for the WCT). By varying the amount of weight in the pack, the degree of incline climbed, and the time engaged in hiking, a firefighter can change the intensity and duration of the workout to meet his or her individual needs.

At times, though, because of weather, terrain availability, or any other number of reasons, it may not be feasible for the firefighter to strap on a pack and head for the hills. If this is the case, options are still available. Walking on an inclined treadmill wearing a weighted vest or pack is undoubtedly the best Plan Bravo. Again, by changing the percentage of incline, the weight carried, and the speed, firefighters can modify both the duration and intensity of the exercise session. Stairmasters™ or other forms of stepmills (supplemented with backpacks, vests, or carrying light dumbbells) can also be used. Hiking on flat ground, be it natural or a treadmill, is obviously not as effective a training surface as inclines, but it is still a very valuable cardiovascular workout, especially when done carrying extra weight.

Jogging, running, cycling, and rowing are also valuable for cardiovascular training but, as we've already said, are not typical activities on the fireline. Again, these forms of exercise can be done, but, for training, more attention needs to be given to hiking with weight, since this best matches our work environment. An exception is if you need to pass a distance run as a condition for employment (for example, smokejumpers). In that case, much more attention needs to be given to running, and training should be focused on running the specified distance within the allotted time.

For those who run as part of their cardiovascular workouts, interval training is an excellent option. Interval training means that the exerciser changes the intensity throughout the workout. Also referred to as Fartlek training (from the Swedish word for "speed play"), a person might run for 10 minutes, walk for 2 minutes, run fast for 2 minutes, then jog slowly for 5 minutes. (Yep, it's called Fartlek. Honest.)

One of the beauties of cardiovascular training is that we do not need to be actually "working out" in order to do it. Many forms of recreation have a built in cardiovascular component. Firefighters tend to be a pretty active bunch. Whether it is rock climbing, downhill- or cross-country skiing, or ball sports, most of us lead active lives. All these activities contribute to our overall level of physical conditioning. The important thing to remember is that some activities are better than others in more closely mimicking the work we do on the fireline. For example, cross-country skiing more closely resembles the demands placed upon the bodies of firefighters than does playing racquetball. Again, these are the types of activities that we need to focus on if our goal is to get in shape for fire season.

In addition to the *level* of exertion, the other question that arises is *how long* should each cardiovascular (CV) training session be? Since the Firefighter Performance Workout has a time limit of between 60-75 minutes, each cardiovascular workout should probably take 30 minutes or less, as time is also needed for warmup, flexibility, core/strength training, and cool down (the other four components of the FPW). So, while 30 minutes is recommended, cardiovascular workouts as short as ten to fifteen minutes can yield significant gains, too. So, even if your workout time is limited, remember that "something" is better than "nothing". Training sessions of 10-15 minutes are less than ideal, but they can be done on an occasional basis. (As they say in winemaking, "don't let *perfection* become the enemy of the *good*".)

Cardiovascular training is one of the most important components of overall physical health. The following list quickly reviews some of the essential points covered in this section of the Firefighter Performance Workout:

- In general, cardiovascular training should be done about 3-4 times per week.

- Pay attention to the intensity and duration of your workouts. Strive for moderation in both of intensity and duration (about 70% to 75% of maximum heart rate and 20-30 minutes per session is our recommendation). Periodically, more intense workouts will need to be done, with higher achieved heart rates and/or longer duration, in order to best mimic those real-world assignments which are extremely demanding physically.

- Focus primarily upon exercises which best mimic our work environment, such as hiking on uneven terrain, or walking on incline treadmills or stepmills, all while carrying a pack or wearing a weighted vest.

- If running is to be done, try some form of interval training or Fartlek training, where you change the intensity of your output throughout the workout.

Core/Strength Training (20-30 minutes)

Strength training is a major component of any effective physical training program. It is also a vital element of the Firefighter Performance Workout. "Strength" is commonly defined as *the ability to exert a force against some form of resistance*, and is generally classified into three areas:

- maximum strength, which is the greatest force that is possible in a single maximum contraction

- elastic strength, which is the ability to overcome a resistance with a fast contraction; and

- strength endurance, or the ability to express force many times over

To varying degrees, each of these three forms of strength is needed in wildland firefighting.

A muscle only grows stronger when it is overloaded, or worked beyond its normal operational level. Therefore, any activity you engage in which overloads the muscles can be considered strength training, be it bucking hay bales or carrying heavy sacks of mail or winter wheat. For many of us, however, overloading of the muscles is achieved through some form of programmatic weight-training, since the person working out in a prescribed training program has much greater control over when they lift, where they lift, and the amount lifted. Weight-training exercises usually involve the use of traditional free weights, a person's own bodyweight, weight machines, medicine balls, or dumbbells. The Firefighter Performance Workout gives the firefighter exercise options for all five of these methods.

As you've probably noticed, this section of the book is called "Core/Strength Training", and this despite the fact that all we've talked about so far is strength training. "Core training" has definitely become the buzz within the field of physical conditioning in recent years, although there tends to be a great deal of misunderstanding about what core training actually means. Contrary to the opinion held by many, working out the "core" involves much more than a few sets of sit-ups. In a nutshell, the core consists of the rectus abdominus, the internal and external obliques, and the erector spinaes. I know, that's a whole lot of Latin to throw into a small English nutshell. Let's translate it to say, these three groups are basically the muscles in your midsection which allow you to transfer movement and power from the upper body to lower body and vice versa. Since so much of what we do as firefighters involves this central portion of the body, the FPW utilizes a considerable amount of core training (not just situps!) in addition to regular strength training.

So, the overloading of the muscles — be they in the "core" or any other part of the body — can be ramped up by progressive increases in:

- the number of repetitions of an exercise

- the number of sets of an exercise, or

- a reduction in the recovery time between sets

Just as with cardiovascular training, core/strength training is often categorized according to intensity — low, moderate, or high. Generally speaking, low intensity usually involves about 12-15 repetitions of a particular exercise, while moderate is in the range of 8-12 reps, and high around 6-8 reps. The higher the intensity of the workout, the more the amount of weight that should be lifted.

Again, it is not the aim of the Firefighter Performance

Workout to prescribe a specific list of exercises that need to be completed. Nor is the FPW regimented with regard to intensity, length of workouts, or frequency. The list of suggested exercises is by no means all-inclusive, either. There are hundreds if not thousands of different exercises, and it is beyond the scope of this book to detail every single one of them. The Firefighter Performance Workout outlined in this book provides suggestions, guidelines, and a number of specific core/strength exercises. Again, the goal is to provide for the firefighter a list from which to pick and choose among the suggested activities, as long as all parts of the body are trained.

The important muscle groups that need to be exercised in every session in order to get a complete full-body workout are the **core**, **legs** (including hips and gluteus maximus or "glutes"), **arms**, **chest**, **shoulders**, and **back**.

The Firefighter Performance Workout in this chapter details two or more exercises for each of these six areas of the body. But again, the FPW allows the person working out to choose and improvise. As a general rule, core/strength training should be done at least twice per week in order to achieve sufficient gains. For those who wish to do more, it is not recommended that core/strength training be done more than six times per week, since it is generally considered best to give the body at least one full day of total rest each week.

CORE TRAINING

Since adequate core muscle development is so crucial to what we do as firefighters, the Firefighter Performance Workout recommends at least two-to-three core training exercises be done during each strength training session. The core can also be worked out by itself on days when no strength training for the other parts of the body is done, since the muscles in the core tend to smaller and recover more quickly.

Start Position

Finish Position

Over the Fence with Dumbbell (DB)

- This is a fundamental exercise for firefighters, since it so closely matches much of what we do on the fireline.

- Feet should be slightly more than shoulder width apart, as above, with toes turned slightly outward as well.

- Start with knees bent, and dumbbell just to the side of one knee

- Lift the dumbbell up and across the body, finishing above the head on the opposite side of body. Return to start position. Repeat.

- Concentrate on rotating at the waist as the weight is being moved across the body.

- Strive for higher repetitions (12-15) per set.

- Rest, then change sides (begin with DB to the side of opposite knee).

- Start with lower weights, because this is more difficult than it looks

- A weight plate or medicine ball can be used in place of the dumbbell, if necessary.

- To make this exercise more difficult, while staying on the ball of your foot, rotate your heel outward as you swing the dumbbell to the opposite shoulder.

Axe Chop with Dumbbell (DB)

- This exercise (see illustrations on p. 41) is also an excellent exercise for the core, as well as several other parts of the body.

- Use the same positioning as in Over the Fence (above): feet slightly more than shoulder width apart, toes turned outward.

- Start with knees bent, and with DB between legs.

- Lift DB upward and over the head while simultaneously straightening legs.

- Return to starting position. Repeat.

Physical Training / 41

Start Position

Finish Position

- Do one set of 12-15 repetitions. This exercise is harder than it looks.

- Substitute a weight plate or medicine ball if no DB is available.

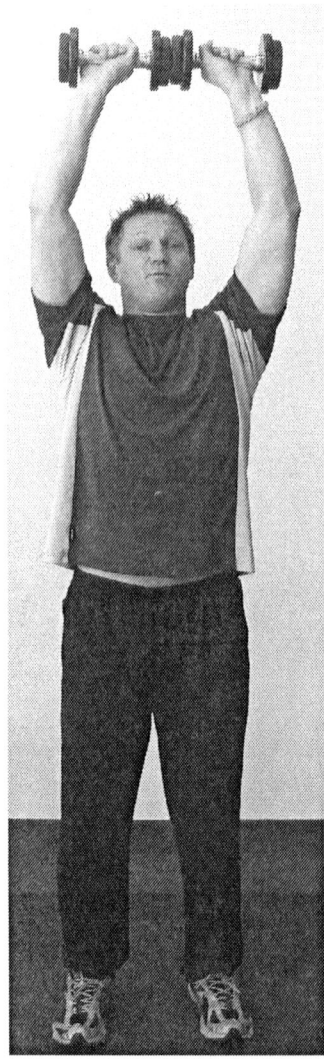
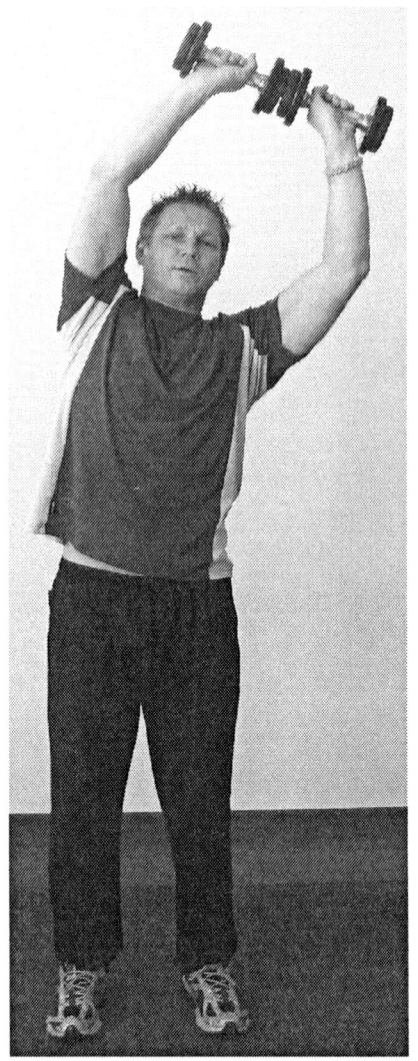

Start Position   Finish Position

Saxon Sidebands

- Start with a light pair of dumbbells in each hand, as shown above (this is harder than it looks).

- Stand with feet shoulder distance apart, knees slightly bent, and arms nearly straight up over your head.

- Stay in a neutral position. Do not lean forward or back.
- Lean as far as comfortable to one side, return to center, then lean to the other side.
- Alternate sides until you have done 10-12 reps for each side.

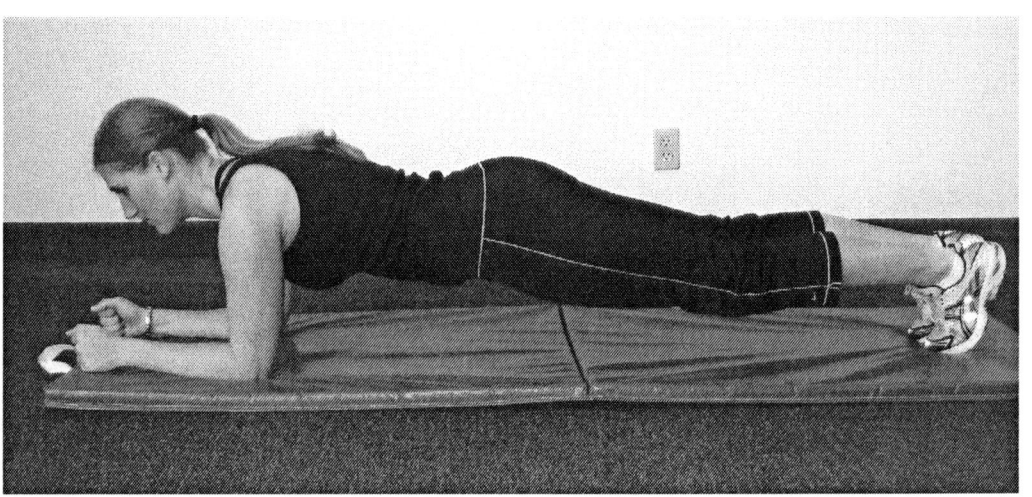

Planks

- Take a prone position with elbows bent, shoulder width apart, and elbows directly underneath the shoulders.
- Feet should be relatively close together.
- Keep chin pointed downwards toward the floor.
- Hold this position for 1-2 minutes. If unable to maintain for this long, reduce time as necessary. Form is much more important than time.
- Repeat after brief rest.
- To make this exercise more difficult, try doing on one arm or one leg.

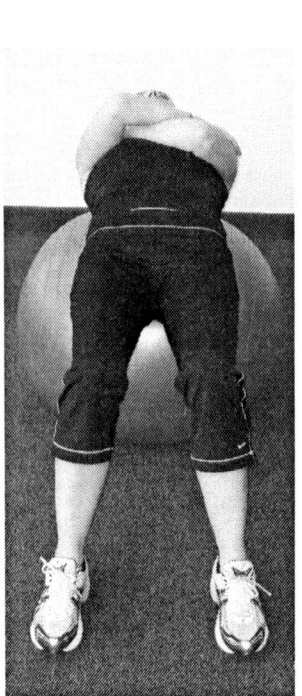 

Start Position   Finish Position

Abdominal Crunch on Stability Ball (SB)

- Slowly roll onto ball until the middle portion of the back is resting on the stability ball. Feet should be slightly narrower than shoulder width.

- Arms should be crossed in front of the body. Do not put your hands behind your head. Keeping chin tucked, do a crunch. Repeat for 20-30 repetitions.

- Experiment with different arm positions (except behind the head) to vary the level of difficulty.

- A weight plate, dumbbell, or medicine ball may be held on chest to increase the difficulty of this exercise.

One-legged row with Dumbbells (DBs)

- This exercise utilizes multiple body parts, with an emphasis on the back and core.

- Begin with one foot on the floor, and the other approximately 2 inches up off the floor and parallel to the one on the floor.

- Start with both of your arms straight and extended toward the floor.

- Execute a rowing action, and then return both arms to the starting position.

- Do 10-12 repetitions of this exercise. Rest, then repeat with the other leg raised off the floor.

- To increase the level of difficulty, progress from both arms to alternating arms. To make it even more difficult, try doing it without shoes, or do the exercise on an unstable surface, such as a core board or Airex pad.

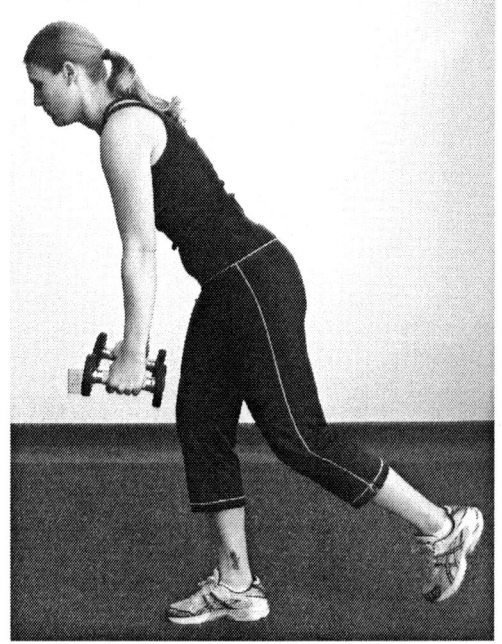

Start Position

Finish Position

LEGS

Obviously, as firefighters we rely significantly on our legs to get us where we need to go and to carry the equipment and supplies needed for our jobs. Legs form the base for all that we do, and therefore must be addressed when doing strength training as well as conditioning training. The legs are comprised of multiple groups of muscles: the quadriceps, hip flexors illiopsoas, adductors, hamstrings, gastrocnemius, soleus, and the tibialis anterior. Together, these muscle groups assist in all the complex movements which the legs are capable of engaging in.

Much of the leg workout comes from the cardiovascular portion of the FPW, especially if one is carrying weight while walking or hiking. That being said, the lower body still needs to be addressed in core/strength training so as to completely work out the entire body. Abdominal bridges are recommended for those who might just be getting started into a core/strength training program, since it is an easier exercise to do. Lunges and deadlifts are somewhat more difficult, so they are recommended for the exerciser who is more advanced in terms of strength and conditioning.

Abdominal Bridge

- Lie on the floor, with legs in the position shown on p. 47.
- Lift the hips off the ground, making sure that the hips stay in line with the back.
- Hold this position to a count of ten.
- Lower slowly back to starting position, and repeat for 12-15 reps.
- Switch legs and repeat.

Physical Training / 47

Start Position

Finish Position

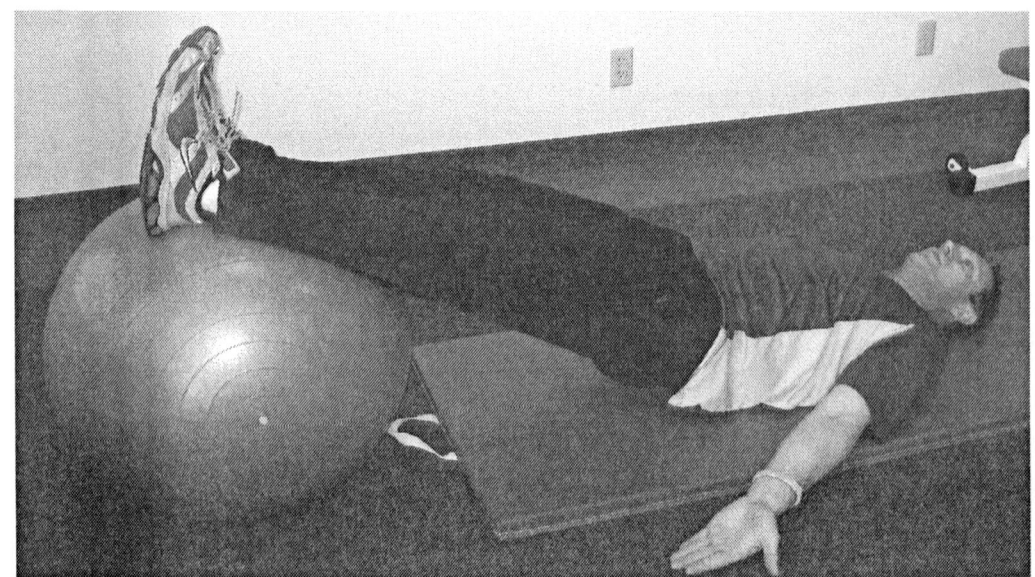

Start Position

Finish Position

Hamstring Rollers with Stability Ball (SB)

- Lie down on a mat or the floor in the position as shown at the left.

- At the start of the exercise, the back of the heels should be on the top of the stability ball, with legs fully extended.

- The hips need to be raised up off the floor, and the back should be kept flat on the floor.

- Hands should be extended out away from the body, with palms facing upwards.

- In a continuous rolling movement, pull the knees toward the chest while keeping the heels on the stability ball.

- Concentrate on keeping the hips raised high during this exercise.

- Slowly roll the stability ball back to the starting position.

- Repeat for 12-15 repetitions.

Forward Lunges

Start Position

- With hands on hips, place feet pointing straight ahead and about shoulder width apart.

- While maintaining optimal spinal alignment, step forward and descend slowly by bending at the hips, knees and ankles. Maintain weight distribution between the heels and the mid-foot during descent.

- Keep the upper torso erect. Do not lean forward. Perform descent slowly. Rise up, and return to starting position.

- Repeat for 12-15 reps. Rest, then repeat on other leg.

- To make the exercise more difficult, hold equally weighted dumbbells in each hand. Experiment with different dumbbells until an optimal weight amount is found.

Finish Position

Deadlift

- This is an excellent lift for the entire body, legs included.

- Begin with hands slightly more than shoulder width apart.

- While bending the knees, lower the bar to a level just higher than the ankles.

- Slowly return to the starting position, and repeat for the desired number of reps.

- Do two sets of this exercise.

- To increase the difficulty, add weight to the bar.

Start Position

Finish Position

BACK

Much of what we do as firefighters (such as digging, lifting, and sawing) involves the back, so it is important to address this area in any strength training program. The back, in basic terms, is a large and complex group of muscles (trapezius, latisimus dorsi, levator scapulae, and rhomboids) that work together to support the spine, help hold the body upright and allow the trunk of the body to move, twist and bend in all directions. Working in conjunction with the core muscles and other parts of the body, the back is a key component in any physical activity. For proof of this, one need look no further than the number of people who experience some form of back pain in their lifetime (50% or more of the general population). While the causes of back pain are complex and multifactorial, effective cardiovascular and strength training can help avoid or reduce its impact on your life and career.

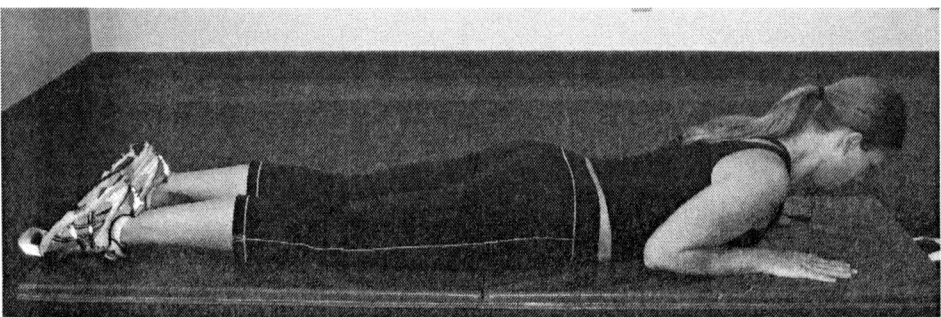

Start Position

McKenzie Press Up

- Avoid this exercise if low back pain is a current issue or if you have had a recent injury to the lower back.

- Lie on your stomach, and place your hands slightly more than shoulder width apart and with the bottom of the palms level with top of shoulders.

- Take a deep breath through the nose.

- As you slowly press upward, exhale through the mouth. Hold this position until it is necessary to take a breath.

- Inhale as you slowly return to the starting position.

- Do ten reps of this exercise.

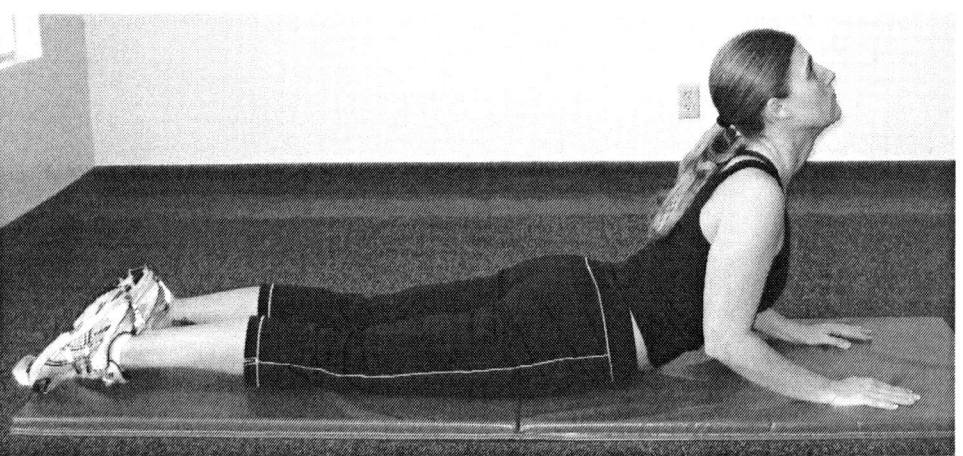

Finish Position

Body Weight Pullups

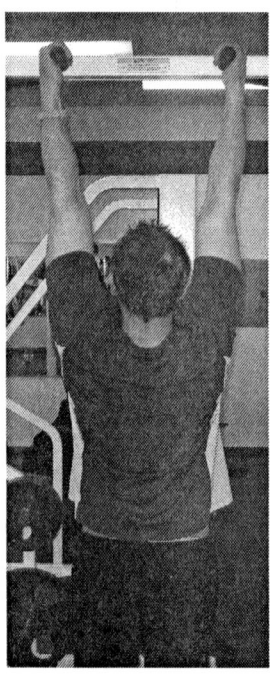

Start Position

Finish Position

- Hands should be placed on the pull-up bar with the palms facing each other.

- Grip should be about shoulder-width apart, with the entire body hanging straight down.

- While maintaining spinal alignment, pull your body upward until your chin is level with the hands.

- Do not "jet" forward with the head, or raise your legs to "kick" your way up.

- Return to the starting position and repeat for the desired number of reps.

- Pull-up assist machines can be used if you are unable to complete enough reps.

- A regular straight bar may also be used if no hand grip bar is available.

Seated Rows

- Sit down on the machine, and make adjustments, if necessary, to obtain proper positioning and weight.

- Hold the bar at chest level with your arms fully extended.

- Flex your elbows and pull your thumbs toward the armpits.

- Hold briefly, then return to the original position.

- Repeat for the desired number of reps.

Start Position

Finish Position

Lat Pull Down with Narrow Grip

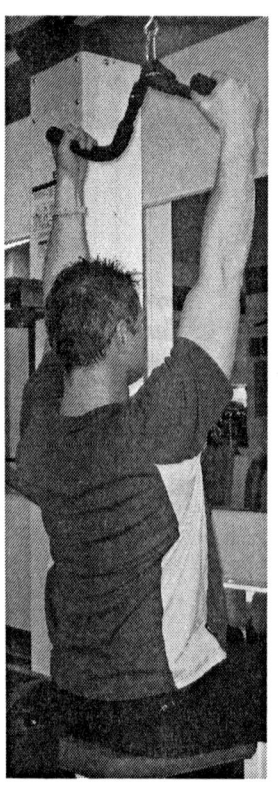

- With palms facing each other, grasp the narrow grip bar.
- Pull the bar down toward the chest by flexing the elbows.
- Do not arch your back or "jet" forward with the head.
- Hold the weight briefly, then slowly return to the starting position by extending the elbows.
- Repeat for the desired number of reps.

Start Position   Finish Position

SHOULDERS

The deltoid is a three-headed muscle that caps each shoulder. These three heads of the deltoid muscle are the anterior, the lateral, and the posterior. All three deltoid heads attach to the humerus, or the bone in the upper arm. The anterior and lateral heads of the deltoid originate on the collar bone (clavicle), while the posterior head originates on the scapula (or shoulder blade).

The main function of the deltoid muscle is to move the arm away from the body, so therefore it comes into play for much of what we do as firefighters. Think about how many times after a long shift of digging a fireline that it is your shoulders that ache the most. The posterior deltoid head helps to move the arm away to the rear, the lateral head away and to the side, and the anterior to the front. Through effective strength training exercises, all ranges of motion for the deltoid can be addressed.

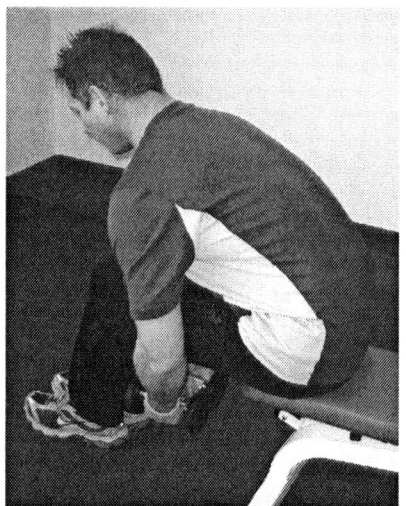
Start Position

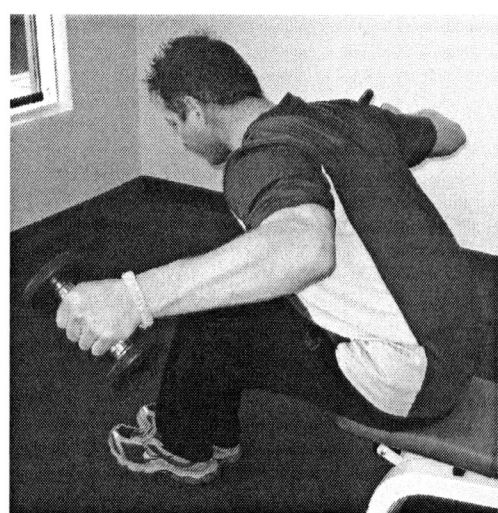
Finish Position

Seated Rear Delt Fly with Dumbbells (DB)

- Sit near the edge of a portable bench, as shown. Hold DBs underneath the legs.

- Bend slightly at the waist, resulting in a "forward lean".

- Keeping elbows bent, pull the shoulders back, ending in the finish position as shown.

- Adjust the weight so that desired number of repetitions can be accomplished.

Around the Worlds

- Begin with feet shoulder-width apart. Hold two light dumbbells in front of you, with palms facing outward, as shown.

Start Position

Finish Position

- Swing dumbbells upward, in an arcing motion, with palms continuing to face outward, until the dumbbells come together over your head.

- Lower the dumbbells slowly to the starting position, and then repeat for 10-12 reps.

- Try to remain erect (standing straight) throughout the exercise. If you find yourself cheating, lower the weight.

Mid-lift Position

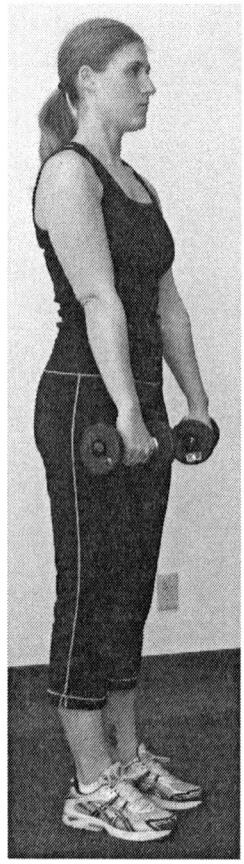 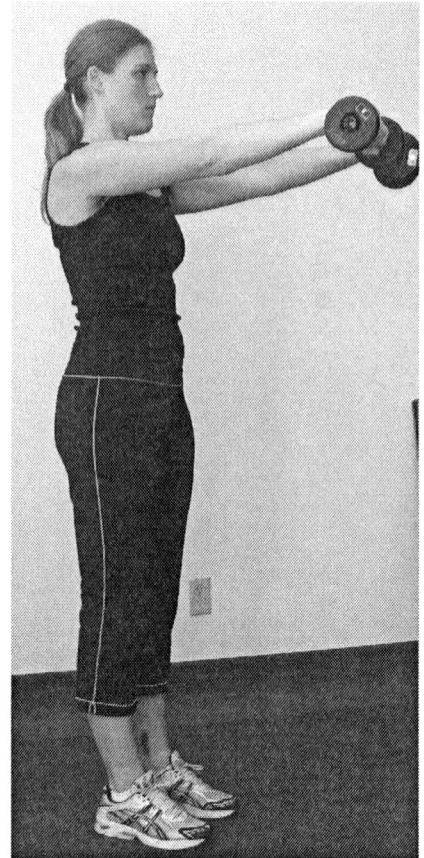

Start Position   Finish Position

Standing Front Raises with Dumbbells (DBs)

- Begin with DBs in front of you, with palms facing inward. Stand upright throughout exercise.

- Raise DBs to about eye level, hold briefly, then return to starting position.

- Repeat for the desired number of reps.

- Begin with lower weight amounts, as this is more difficult than it looks.

Upright Row with Barbell

Start Position

Finish Position

- Begin with the feet about shoulder-width apart, and grasp the bar with hands placed about shoulder-width apart as well.

- Bring the bar up to about the level of the sternum (the bone at the center front of the chest connecting the two halves of the ribcage).

- Lower the bar slowly to the starting position, and repeat for the desired number of reps.

- Add weights to the bar to increase the level of difficulty.

62 / Part One: Physical Factors in Optimal Performance

Start Position

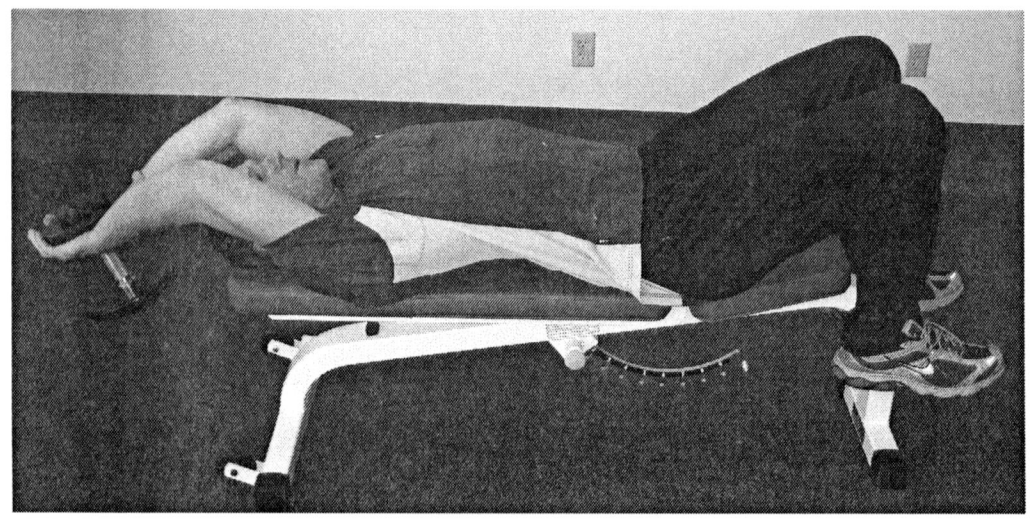

Finish Position

Pullover on Bench with Dumbbell (DB)

- Lie down on the bench, grasping the dumbbell as shown on page 62.

- Slowly lower the dumbbell to a point where the arms are in alignment with the back, and nearly parallel to the ground (or to a level that is comfortably within your range of motion).

- Pull the weight back to the starting position, and repeat for the desired number of reps.

CHEST

The chest consists of two muscles, one large and one small, named the pectoralis major and the pectoralis minor respectively. The pectoralis major (or pecs) are located on the front of the rib cage. Initiating on the breast bone in the center of the chest, they attach to the shoulder joint near the humerus. The muscle fibers of the pecs fan out across the chest, thus allowing the humerus to move in a multitude of planes across the body.

Located underneath the pectoralis major, the pectoralis minor attaches to the coracoid process of the scapula, and originates on the middle ribs. It helps to move the shoulder area forward, such as when you shrug your shoulder in a forward motion.

Since the chest muscles are so involved in arm and shoulder movements, they too must be addressed during workouts. Plus, nice pecs look good at the beach!

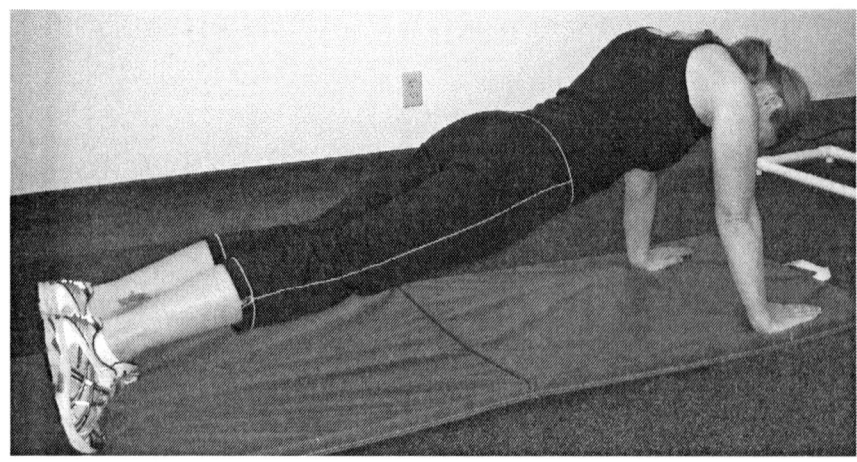

Start Position

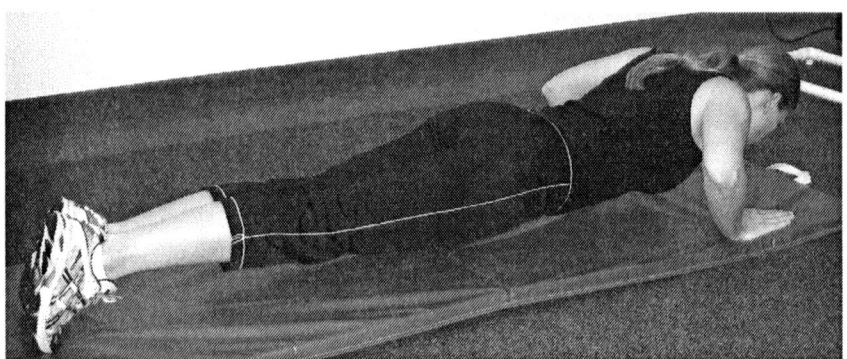

Finish Position

Pushups

- Assume the starting position as shown. Begin with your hands outside the width of the shoulders.

- Lower yourself until your elbows are bent at 90 degrees. Hold briefly, then press back up to starting position.

- Repeat for 15-20 reps, or until you cannot continue. To increase the difficulty, slow down on the descent (lower to a count of 4, for example).

Start Position   Finish Position

Seated Chest Press

- Before starting, adjust the height of the seat so that the machine handles are in a line with a point just below the bottom of your shoulders.

- Grip width can be anywhere on the handles that feels comfortable to you. The width of grip will have some impact on the area of the chest that is worked out. Therefore, if multiple sets are done, each set should be with a slightly different grip width.

- Concentrate on a slow and smooth range of motion throughout the lift.

- The number of repetitions is up to you, depending on the level of intensity you are after.

Start Position

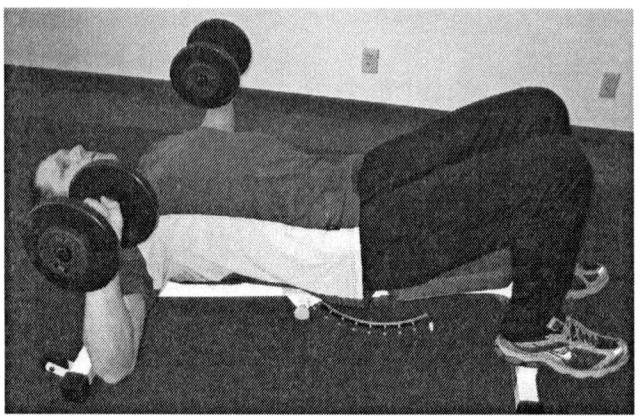

Finish Position

Chest Press with Dumbbells (DB)

- Lie down on the bench as shown, with DB raised and arms extended but not locked out.

- Feet should be pointing straight ahead, and back should be flat on the bench.

- Lower the weight slowly, hold briefly at the bottom of the motion, then lift it back up to the starting position.

- Remember not to arch the back at any time during lift.

Arms

While many people believe that the arm is composed of only the biceps and the triceps, it is actually made up of four distinct muscles.

- The biceps, a two-headed muscle located at the front of the upper arm, is responsible for flexing and curling the lower arm (below the elbow) relative to the upper arm. It is also involved in those motions where the forearm is twisted upward.

- The brachialis is a muscle that covers the front of the elbow, and it comes into play when flexion occurs at the elbow joint.

- The brachioradialis (or forearm flexors) is a muscle located on the outer, or radial, side of the forearm, which is responsible for downward and upward twisting of the forearm.

- Lastly, the triceps is a three-headed muscle situated at the rear of the upper arm. It is involved with any straightening (or extension) of the arm.

When working out the arm, two different lifts (at the very least) need to be accomplished: one lift that addresses the curling motion of the arm, and one lift that focuses on the extension of the arm. A third lift, one that specifically works out the forearm flexors, can also be incorporated, since the use of many fireline handtools and chainsaws involves this brachioradialis muscle.

Start Position

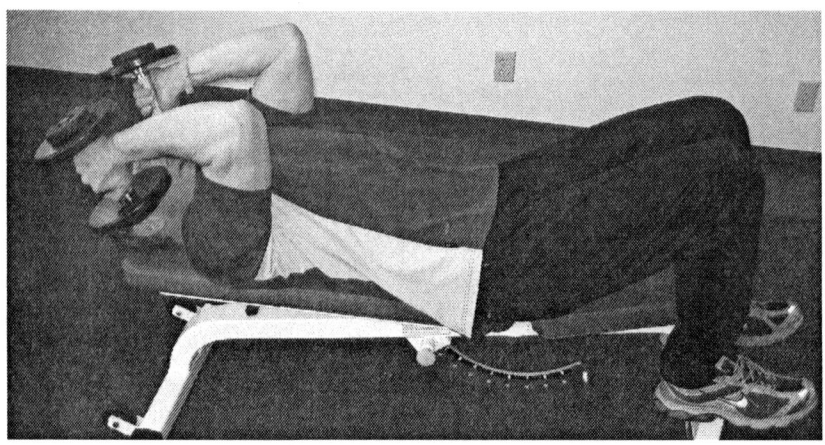

Finish Position

Skull Crushers with Dumbbell (DB)

- While sitting on bench, grasp a DB in each hand. Lie down on your back in the starting position, as shown.

- Slowly lower each DB to a point just above your forehead. Focus on bending your arm at the elbow only.

- Slowly return the DB to the starting position. Repeat these steps for the desired number of repetitions.

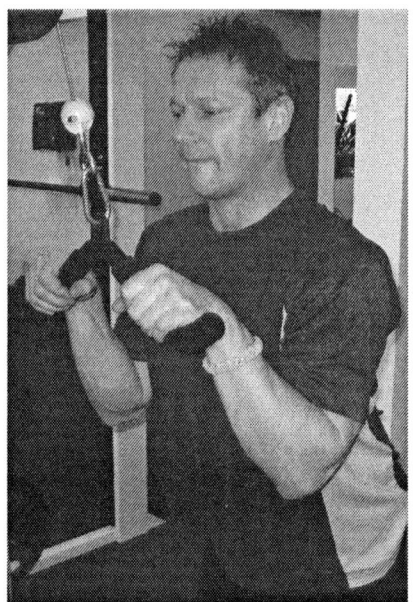
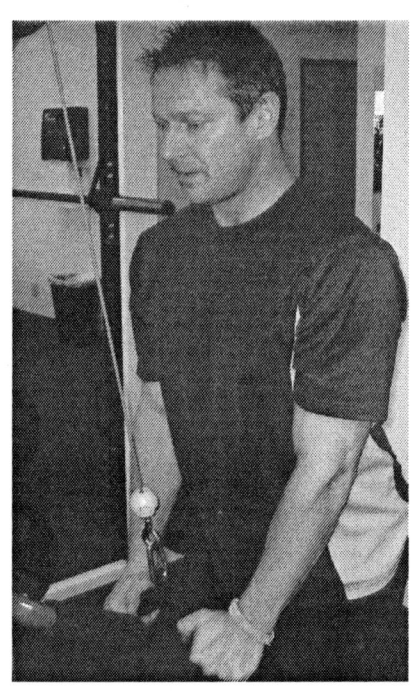

Start Position (above); Finish Position (right)

Standing Triceps Extension

- Feet and hands should both be about shoulder-width apart, and knees slightly bent to ensure a stable stance.

- Begin with the weight at about chest level. Maintain shoulder position, and extend the forearms downward.

- Hold briefly at the bottom of this motion, then slowly raise the arms back to the starting position.

- Keep focus on the triceps by maintaining postural alignment throughout the exercise. Repeat steps as desired.

Start Position

Bench Dips

- This is a good exercise not only for the triceps, but also for the chest and components of the shoulder.

- Assume a beginning position as shown. The feet should be close together, and hands should be close to the sides of the body.

Finish Position

- Slowly bend the arms, and lower your body down to the point where the elbows are just below the top of the shoulders. Inhale on the downward motion, and exhale on the upward return to the original starting position.

- Concentrate on slow and smooth motions. Lower your body only to a point where it is still comfortable to do so.

- The number of reps is up to you, but shoot for a higher number that fatigues the muscles.

Start Position

Standing Biceps Curl with Dumbbell (DB)

- Grab a DB with each hand.

- Feet should be about shoulder-width apart, with toes pointing forward, and knees slightly bent.

- Begin with DB as shown, and perform a bicep curl.

Finish Position

- Lower the weight slowly back to the original starting position.
- Experiment with different weights until you find one that allows you to do at least 10 repetitions, but still leaves you straining on the last few reps.
- Kiss your biceps!

Some important points to remember concerning core/strength training are:

- Core/strength training should be done at least twice per week, but no more than six times per week. The body needs time to rest and recover, so avoid over-training!

- Each of the six areas (core, legs, arms, chest, shoulders, and back) need to be worked out during each core/strength training session. I like to use the acronym **CLACSB** to remember all of them. Since we work in fire, most of us are used to acronyms, right? What could be more memorable than **CLACSB**?

- Stay within the 20-30 minute time-frame for your core/strength training workouts. As a suggestion, do one set of one specific exercise in each of the six areas. A sample workout might be as follows:

- Core: 35 medicine-ball sit-ups

- Legs: one set (12 reps per leg) of forward lunges

- Arms: one set (12 reps) of bicep curls

- Chest: one set (10 reps) of chest press with dumbbells

- Shoulders: one set (10 reps) of upright row with barbell

- Back: one set (12 reps) of seated rows

After you have covered all six areas, with any time that remains you can go back and do a second set of each exercise, or pick a new exercise for each of the six areas of the body (CLACSB) and complete one set for each of those.

Here are some more pointers to assist you:

- Vary the intensity of your workouts by alternating the amount of weight lifted and the number of repetitions.

Some days you should work out with lower reps and heavier weights, and then some days with higher reps and lighter weights, and then some days for variety you can shoot for something in the middle for both reps and weight.

- Strive for "constant tension" in all of your lifts. Do not rest weights on your body (such as letting the bar sit on your chest when doing the bench press). Try not to lock out any joint when lifting. Rather, stop at a point just before locking out (for example, when doing dips, return to a starting position where the arm is still slightly bent, not straight.) Remember to inhale at the beginning of lifts, and to exhale slowly when exerting the weight.

- Be efficient as you transition between the lifts, as this will save you time. Minimize the distance you have to walk between stations. Try to have as many of the weights (dumbbells, barbells, machines, medicine balls) as close to you as possible, so that when you finish one exercise, you can quickly begin another. This saves time in your workout and also keeps your heart beating at a higher rate, which is good for your cardiovascular system.

Cool Down (5-10 minutes)

The fifth and final component of the Firefighter Performance Workout is a recovery period after the workout has been accomplished. This period of time is referred to as "the cool down." While some believe the purpose of a cooling off period is to prevent muscle soreness from developing, a cool down actually helps to prevent or minimize dizziness after a vigorous workout.

During exercise, the muscles of the body serve a pump-like function, as they help to move blood through the body and back to the heart. When muscles are in a relaxed state, the veins that are located near them begin to fill with blood. When these same muscles are contracted, these veins become squeezed, thus helping to pump blood back toward the heart. Should a person stop exercising suddenly after an intense bout of exercise, these muscles stop "pushing" blood back to the heart and it pools up in the exerted muscle(s). The heart now has to take on all of these pumping duties by itself. If this happens, dizziness can often times result because sufficient blood flow is not being delivered to the brain. Result: major head rush, dude!

Cool downs can be accomplished in a couple of different ways:

- By continuing with your present exercise, but gradually slowing its intensity. For example, if you are finished doing Seated Chest Presses with dumbbells, you can slow the intensity by doing another set of the same lift, but with much lower weight and a lower number of repetitions.

- Through the use of slow jogging or brisk walking. If you just finished a demanding 30 minute trail hike with a heavy pack, take the pack off and simply walk around on flat ground for 5-10 minutes.

The key with cool downs is that they do not have to be very long, which fits in nicely with the overall time structure of the Firefighter Performance Workout.

Should time allow, a quick set of stretches can also be incorporated into your cooling down period, since your muscles are still warm at this time. Stretching, as noted earlier in the flexibility section of the Firefighter Performance Workout, has multiple benefits: it relaxes the muscles, it helps to restore them to their resting length, and it improves flexibility.

The FPW in Review

Utilizing five unique, separate components (**warmup, flexibility, cardiovascular training, core/strength training,** and **cool down**), the FPW meets several needs of the wildland firefighter. To begin with, it is an all-body workout, which is important since firefighting is an all-body job. Secondly, building on the fact that firefighters are unique and individualistic at times, the FPW allows each person to tailor their workout to their own needs while keeping within the FPW's overall framework. And finally, the FPW is a time-limited workout program, since for most of us, our physical training time (like the rest of our time) is often in short supply. Therefore, in a 60-75 minute period, the entire workout can be accomplished.

Again, in order to best match the fire environment and to "train like you fight," the **cardiovascular** and **core/strength training** sessions should be alternated when using the FPW. For example, sometimes we have long, difficult hikes into our incidents before we have to start digging line, dragging hose, or running saws. When this happens, we engage our cardiovascular systems before we rely on our core/strength systems. At other times, we helitack, jump, or even drive right up to the incident, and in so doing place intense demands upon the core/strength muscles long before fully utilizing our cardiovascular systems.

By varying the order in which we complete these two components of the FPW, our workouts will better replicate our actual work situations. All other parts of the FPW should be completed in the order given. **Warmup** needs to be the first activity, with **flexibility** coming second. **Cardiovascular** and **core/strength training** sessions can be alternated every workout, every week, or in whatever way you decide works best for you. **Cool down**, regardless of the order in which the cardio and strength training units is completed, should always come as the last activity done in the FPW.

At least once every two weeks, you should have a workout that, although it might not leave you crying for your primary caregiver, really kicks your butt. Push yourself hard during this session! Remember, you have to train like you fight. Every now and then we have fires which really put us to the test physically. If you have really challenged yourself during some of your workouts, your body will be better prepared to meet the demands placed upon it by one of these nasty blazes.

The question that often arises regarding a workout program has to do with how soon a person should start one prior to their official "season." From a broader time perspective, most physiologists and exercise professionals agree that about four months of physical training are needed to prepare the body for vigorous work and exercise. Luckily, most firefighters stay active throughout the year. Therefore, they are seldom "starting from scratch" when it comes to getting into shape.

The Firefighter Performance Workout can be started at any point, be it the winter "slow time" or smack dab in the middle of fire season. Obviously, once the season has begun, physical training times become less consistent, especially during busy fire years. This will make it more challenging to get consistent workouts accomplished. However, busy fire seasons usually result in peak physical conditioning, since we are consistently placing heavy demands on our bodies, day after day. Bottom line: you can start the FPW whenever you like, but just make sure you do! The FPW program provides a complete full body workout while simultaneously yielding the most bang for the buck as far as time efficiency goes. Definitely a win/win scenario.

Nutrition, Hydration, and Ergogenic Aids

August, 2000: We jumped just after dawn, while the air was still calm and before the enemy winds that had blown all summer had a chance to build. The jumpspot consisted of a small clearing on a steep hillside in Idaho's Panhandle National Forest (IPF). The IPF had been a hotspot for fires for two weeks, and many of us had fought small blazes there just a few days before. It had been so busy here, in recent weeks, that we had begun to refer to ourselves humorously as "panhandlers" due to our extended stays. Although we had no way of knowing, it would become a moniker we'd use for weeks to come. Before the summer was through, most of us would be back in these parts fighting fires again.

By this point in the busy fire season of 2000, the smokejumping process had become a rather well-rehearsed ballet. After the jumpspot was selected, paper streamers were thrown out at 1500 feet above the ground to determine wind direction and speed. "It looks like about a hundred yards of drift," the spotter yelled above the windblast of the open door. Sarah, the jumper in charge of our group, nodded her head and tightened the straps on her harness. The Incident Commander on the fire below, himself a smokejumper, had called for a reinforcement of eight more jumpers. Looking at the ten acres of active fire from the plane, we knew that we would have our hands full. After the raw thrill of the jump had worn off, what lay ahead of us would be long days of digging fireline and more of the same bland food that we had been surviving on all summer.

In groups of two we leapt from Jumper 15, a venerable DC-3 which had come off the production line well before the ink was dry on the treaty ending World War II. Jumping in the second group of two, I listened attentively to the spotter's briefing. "The first two made it in there just fine. Good luck, and see you in a few days!" I nodded my head to signify that I understood, and assumed my position in the open door. "We're on final. Get ready!" I tensed in anticipation of the slap on my calf that would be the signal for me to jump. When it came, I launched myself into the brisk morning air over the state of Idaho.

"One thousand, two thousand, three thousand, four thousand, look thousand!" I yelled as my body tumbled away from the plane. Reaching the end of my count, I looked above to see the beautiful blue and white canopy from which I was now dangling. I reached for my steering toggles, and craned my head to find my jump partner who had also just completed his own leap of faith. Seeing that we had good separation, I turned to face the wind. The clearing was behind me now. I crabbed back and forth across the windline trying to figure out my approach. Since the parachute ride from sky to earth takes only about 90 seconds, my actions had to be precise. Quickly, instinctively, I calculated my position in reference to the jumpspot. I came in from the uphill side, thus giving myself room to fly downhill should I need the extra landing space. Hitting the ground, I rolled down the slope and skidded to a halt. I unbuckled my parachute, and said a quick prayer of thanks to Big Ernie, the god of Fire. I looked skyward for the next group of jumpers. Two by two, the rest of the animals left the ark in search of solid ground.

When all had made it safely to terra firma, we sacked up our equipment and toted the sixty-pound bags to the top of the hill. Mark, the Incident Commander on the fire, was there to greet us as we crested the ridge. He had jumped the fire late yesterday afternoon, but the winds had been squirrelly and his trip to the ground had not been a pleasant one. He recounted that at one point the wind blew so hard that both he and his parachute were on the same level plane. An important rule of thumb in smokejumping entails always trying to beat your parachute to the ground. Mark had done his best to heed this simple advice, but the gusting winds apparently had some of their own ideas. Once on the ground, he called the plane on his radio and cancelled the remaining jumpers. However, there was one small catch. Since jumpers always work with a minimum group size of two, somebody else needed to jump to complete the pair. This task fell to the next person on the jumplist, a rookie by the name of Derek. On our first night, around a dying campfire, Derek would share the details of his own harrowing jump. "No kiddin', I was being blown faster backwards then I've ever flown going forward," he would tell us, speaking as much to the burning embers as to those sitting around them. Somewhere in the process, two silver dollars had taken up residence in the place his eyes used to call home, and it looked like they planned on staying awhile.

With Mark leading the way, our group walked along the spine of the ridge in the direction of the billowing smoke. Our cargo had been thrown into a large rockslide approximately half the distance between our jumpspot and the fire. Cardboard boxes, ranging in weight from fifty to one hundred pounds, lay scattered below the ridge. One by one, we carted them up the hill, stopping often to cool the burning in our lungs. Inside these boxes were the tools, water, and food that had sustained us all summer, and would continue to do so on this newest blaze. As far as I was concerned, though, the tools and water would have been enough. It was not that the food itself was that bad. Rather, with the heavy activity of the fire season, I was becoming tired of eating the same things over and over. It had become a sort of culinary version of the movie "Groundhog Day," with each day bringing the same meals as the day before.

The packages had stamped on them an assortment of colorful names: Leonardo de Fettuccini, Santa Fe Chicken and Rice, Honey-lime Chicken, or Albacore Tuna with Buttery Noodles. By this point, though, I was convinced that it was all a cruel joke from the *factory. While the adjective-laden names on the outsides of the packages might differ, I was sure that the contents all came from the same vat. I was sick of adding boiling water to everything that I consumed. Hot water for my coffee, hot water for my oatmeal, hot water for my freeze-dried chili, hot water for my rice. I longed for anything that did not have, "Add hot water" as the first step in the directions for preparation. Digging through the food boxes, I filled my pack with more of the same, and resigned myself to the prospect of eating perpetual re-runs.*

Within minutes we had reached the fire. Mark quickly lined us out, and we went to work. At a size of about ten acres and with only ten smokejumpers on the scene, even I could figure out the math. It equated to an acre per person. If the wind started blowing, like it had been doing all season, the math would become more complicated. While the number of jumpers might stay constant, the fire would be free to practice its own version of the multiplication tables. The chainsaw teams went first, cutting the brush that was ten feet high in places. I grabbed a Pulaski and fell in line with the other diggers.

Progress was maddeningly slow. The sawyers were faced with a smothering blanket of green brush. The swampers, those responsible for moving and stacking what the sawyers cut, had perhaps the most difficult job. Scratched and bleeding from their battles in the brush wars, they trudged on. We diggers did not fare much better. Below the brush lay a dense mat of roots, rocks, and thick bear grass. Our tools, dulled within minutes, sent sparks flying each time the blade hit a rock on the downward strike. Hands and minds became numb with the repetition. Hours later, when word came down the line to take a break for lunch, I dropped where I stood. A combination of heavy exertion and the thought of eating one of the same old fares had eradicated my appetite. I slugged a quart of water and opted for a quick combat nap in the dirt and rocks instead. I awakened when the sawyers fired up their Stihl's, and I groggily reached for my Pulaski and resumed digging.

For the rest of the afternoon and into the evening, we continued cutting our line around the fire, trying to starve it of the combustible material it needed to exist. In their own way, ironically, the fires of 2000 were doing to me the very same thing I was doing to them. By skipping meal times, I was starving my own body of the fuel it needed to continue functioning. Something had to give. Around 2100 hours, we headed back to our makeshift camp on the top of the ridge. Despite the long day of work, the fire was still not even halfway lined. Water was placed on the cooking fires to boil, and the others began to pick out their evening meals. I snarfed down a melted candy bar, and brewed myself a cup of coffee. Knowing that I needed to eat something more, I dug for the rice at the bottom of my pack. It would be bland, but I preferred that over one of the freeze-dried meals with a fancy name. Later, with dinner under the belt, the crew migrated to various parts of the fire to check its status. After a couple hours of close monitoring, we decided to call it a night. I found a relatively flat spot amidst the rocks and tree litter, and rolled out my sleeping bag. On yet another night, on yet another fire, I fell asleep with an insatiable knot in my stomach.

The sawing and digging continued for most of the next two days. By the middle of the afternoon on the third day, we had at last accomplished what jumpers refer to as "driving the golden spike." This occurs when the last slice of earth separating the beginning point of the fireline from the ending point has been removed. With the fire fully lined we could breathe a small sigh of relief. Combining equal parts hard work and luck in the form of windless days, it looked as if we would catch another fire before it gobbled up the Idaho countryside. However, our group was still facing the reality of several more days of mop-up to ensure the fire was dead out.

Up until now, I had been eating as little as possible, subsisting mostly on rice, or what I had dubbed the breakfast, lunch, and dinner of champions. Then, without forewarning, the leaves in the trees began to rustle and the winds of change moved through our camp. We all sat staring at the cooking fire, and I sensed the light breeze as it touched my exposed skin. Our culinary fortunes were about to improve. Mark, the Incident Commander, was the first to speak. "You know, we're gonna be here for several more days mopping up. How 'bout a fresh food drop? Anybody else up for that?" I felt like kissing him on the lips, but stifled the urge for fear that my excitement over fresh food would be mistaken for something else. In unison, the nine of us voiced our approval.

Fresh food drops are a coveted amenity in the world of smokejumping. They occur rarely, and when one does take place it is not soon forgotten by those fortunate enough to be a part of it. Ice chests, stuffed with steaks, and pancake batter, and eggs, and milk, and bread, and cereal, and cookies, and goodies too numerous to mention, are fitted with parachutes and dropped from the same planes used by the jumpers. Fresh food drops are like the star player of your favorite team, often able to change attitudes and outcomes by their mere presence. "I'll order it tonight, and it should be here sometime tomorrow," Mark announced to the salivating masses. While lying in my sleeping bag later that night, the hands of expectation loosened slightly the knot of hunger that had formed in my gut.

By mid-morning, the Sandpoint dispatch center had relayed to Mark that the drop would occur at 1800 hours. We mopped up through the day, compulsively checking our watches at every opportunity. "Fresh food in four hours," someone would whisper in a hushed tone to the person next to them, as if afraid of letting the rest of the world in on a secret. Time moved in imperceptible bits, and we continued mopping. 1800 hours came and went. Around 1900, the heavy drone of the DC-3's engines could be heard as it approached from the east. As if on cue, twenty eyes looked skyward and ten stomachs rumbled. "We've got six total boxes, for a total of eleven hundred pounds," the spotter announced over the radio. "We'll be throwing them one at a time, trying to put them on the top of the ridge." Eleven hundred pounds, I thought to myself. Even subtracting out the weight of the coolers and the ice inside them, that still left several hundred pounds of food. Christmas had come in August this year.

But now, as if in spite, the wind that heretofore had been calm began to gust. The DC-3 rocked as it made its approach. The first box was kicked out just over the ridge, and for a few tantalizing seconds it looked as if it might land somewhere near the top. Then, the winds overpowered the canopy, pushing it farther and farther down the steep hill on the northern edge of the fire. Upon impact, it disappeared into a sea of thick brush. We watched in silence as the remaining five boxes mimicked the actions of the first. Our fresh food was not to be had without a price. The boxes, nearly two hundred pounds apiece, would have to be hauled up to camp at the top of the ridge.

We split up into groups and went about searching for the canopies and the precious cargo attached to them. "Here's one!" a voice exclaimed from the sea of brush. "We've got another one over here," the brush replied from farther down the hill. With jumpers on each end of the boxes, they were pushed and drug from the heavy green foliage to the main fireline. From there, they would have to be carried to the top, a distance of several hundred feet, with an incline approaching nearly eighty percent. An hour later, all six boxes had been found and moved to the staging area. Inch by excruciating inch, the cargo began its journey to the ridge.

At that moment, we had become anachronisms in time. Although the 21st Century had dawned for others, somehow we had regressed to the stage of primal humans. Disheveled and unkempt, sleeping in the dirt, and battling the elements of earth and fire, our nomadic band struggled to carry the food that would keep us alive. With strained muscles and rivulets of sweat pouring down our faces, we moved closer and closer to the top. Minutes crept by at a pace matched only by that of the cargo. "Maybe it'd be better if we just left it and kept eating freeze-drieds," someone quipped. Subdued laughter passed down the line, but everyone kept on heaving. With a final mighty push, we crested the top. It was now only a short drag into the main camp.

Once there, the boxes were attacked by the famished mob. They had come filled with everything we had anticipated, and much more. Those things that did not need to be kept cold were removed and organized into an open-air pantry. Meats, cheeses, milk, and other items needing ice were kept in their coolers. Cups, plates, utensils, and other cooking tools were pulled aside and placed in a makeshift kitchen area. After the organization process had been completed, the feasting began. Fires were stoked. Steaks as thick as a man's arm were unwrapped and thrown onto cooking grills. Gargantuan potatoes were sliced into chunks, wrapped in tinfoil, and thrown into the coals. Cookies were inhaled by the handful, no one worrying in the least that their appetite might be spoiled by doing so. Fresh vegetables, hamburgers, tortilla chips, gallons of grape juice, crackers, it did not matter. All became quick casualties in the war against hunger. I ate for hours, anything and everything I could get my hands on, until at last exhaustion overpowered me. I passed out like a drunk.

For three more days we mopped up, meticulously snuffing out every hot spot. With the fresh food, spirits and attitudes soared. Each morning we stuffed our packs with thick, meaty sandwiches. Canteens were filled with cold milk, and rolls of Oreos were safely tucked away in our bags so as to not be crushed. In the evening, when the day's work was done, the eating continued, often times until the transformation of a new day. On our last morning, the rookies were put in charge of making brunch. While the rest of us tended to the fire, they prepared a feast of epic proportions. When they called over the radio to tell us that it was ready, we double-timed it back to camp. Freshly picked huckleberries had been added to batter, and hubcap sized pancakes were cooking in iron skillets. Eggs, toast, bacon, sausage, ham, and fruit salads rounded out the menu. We ate until there was nothing left.

The summer of 2000 will long be remembered for its fires and the intensity with which they burned. In their aftermath, they have left ashes, charred trees, and a landscape forever changed. Also left behind have been lessons, lessons now indelibly etched into our memories. We learned that despite our best efforts to control the burning world around us, sometimes nature has a brutal and devastating way of reminding us that at times we are mere spectators to her fury. We learned once again that when the chips are down, we are at our best when we collectively work together. And lastly, and perhaps most usefully, we re-learned what children have known for decades. That lesson, beautiful in its simplicity, is that a handful of cookies and a quart of cold milk can go a long way toward changing our disposition.

Nutrition

If the body can be compared to the engine of an automobile, then food serves as the fuel that powers this engine. Everyone has heard the saying, "You are what you eat." Well, if that is the case, then I fully expect to become a chocolate-covered doughnut in the very near future. Let's just say that I don't exactly follow the "my body is a temple" philosophy at all times. As a firefighter, I have consumed way more than my fair share of MRE's and Rat Packs. On initial attack assignments, I must admit that the candy bars are the first things eaten when I rummage through the food boxes. Since I do not care for freeze-dried meals, I sometimes get by on as little food as possible. In fire camps with caterers, my focus is usually on the dessert bar, even though I haven't necessarily "cleaned my plate" beforehand.

I'm smart enough to know this approach to personal nutrition is not the wisest move. I often rationalize such eating behavior by telling myself that I am working very hard, and therefore burning off all the extra calories from the "sweets." However, I also admit that earlier in my career I was just not very knowledgeable when it came to what was best to eat in order to power my internal "engine."

In such ignorance, I was not alone. In a survey of 123 wildland firefighters, Kodeski and others (2004) found that fire crew members had numerous misconceptions concerning the nutritional demands of wildland firefighting. Among other things, respondents were shown to be unaware of the role of carbohydrate, fat, and protein as energy sources during arduous work. The contributions of vitamins and minerals were also unclear.

Fighting fire can be extremely demanding on the body in terms of nutritional demands. Research by Brent Ruby and others at the University of Montana Performance Laboratory shows that some male firefighters may require more than 6000 kilocalories of food daily, while some females may need more than 5000 kilocalories over the course of a long, hard shift. For reference, calories on a food package are actually kilocalories (1,000 calories = 1 kilocalorie). Therefore, a can of soda which lists 150 calories on the label actually contains 150,000 regular calories, or 150 kilocalories (kcals).

In this same field-based research by Ruby and others, male firefighters burned an average of 4758 kilocalories a day while they consumed only 4068 kilocalories. This resulted in a daily caloric deficit of 690 kcals, and led to a 1.25 kilogram (2.76 pound) weight loss over the course of a 14-day assignment.

Female firefighters in the same study burned an average of 3550 kilocalories per day, while consuming 3222 kcals. The daily caloric deficit in this case was an average of 328 kcals, which led to an average loss of .6 kilograms (1.3 pounds) during a 14-day assignment.

Obviously, when these types of deficits occur, something has to make up the difference. If energy intake falls short of what the body is burning, eventually the body may begin to derive energy from muscle tissue. Put another way, the body starts eating itself. Such loss of lean tissue increases the likelihood of fatigue and reduces the ability of a firefighter to perform difficult tasks inherent in wildland firefighting. When deficits of this nature occur frequently, as is the case during especially busy periods of the fire season, the resulting loss of body weight and lean body mass can become particularly troublesome.

The need for a diet high in carbohydrates and low in fat is supported by studies of endurance athletes, soldiers, and wildland firefighters. The recommended diet for these high-octane groups is often referred to as The Performance Diet. Research in support of The Performance Diet for wildland firefighters can be found in the Fall 2002 *Wildland Firefighter Health and Safety Report* by Dr. Brian Sharkey, project leader at Missoula Technology and Development Center and professor emeritus of the University of Montana Human Performance Lab, and others.

The Performance Diet

Nutrient	Percent of Daily Calories
Carbohydrates	60
Fats	25
Protein	15

Carbohydrates are broadly defined as any of a group of organic compounds that include sugars, starches, celluloses, and gums that serve as a major energy source in the diet of animals. These compounds are produced by photosynthetic plants and contain carbon, hydrogen, and oxygen, usually in the ratio 1:2:1. Complex carbohydrates (such as whole-grained breads and pastas, potatoes, corn, rice, and beans) provide energy and also supply needed nutrients and fiber.

Fats, one of three main food forms needed by the body, are the main energy store in the body; they also act as insulating material under the skin and around some organs. Fats come in two forms: saturated and unsaturated. Without getting into complex biological processes or definitions, unsaturated fats are generally regarded as healthier in the diet than saturated fats.

Proteins come from any of a group of complex organic macromolecules that contain carbon, hydrogen, oxygen, nitrogen, and usually sulfur and are composed of one or more chains of amino acids. Proteins are fundamental components of all living cells and include many substances, such as enzymes, hormones, and antibodies, that are necessary for the proper functioning of an organism. They are essential in the diet of animals for the growth and repair of tissue and can be obtained from foods such as meat, fish, eggs, milk, and legumes.

In order to achieve target percentages of the Performance Diet (60% of calories from carbohydrates, 25% from fats, 15% from protein), firefighters must pay attention to what they eat, and follow nutritional information included on the labels of most foods. Paying attention to labels can be difficult, especially since firefighters must often rely on others to bring their food to them. Computer analysis of food provided in fire camps indicates that most meals exceed dietary recommendations for fat and are low in nutritional complex carbohydrates. To address this imbalance, as firefighters we need to get informed and then take a more active role in choosing what we eat and what we do not eat.

Sharkey and others (2002) suggest that a firefighter can approach the High Performance diet percentages by:

- Increasing the intake of complex carbohydrates (potatoes, corn, rice, beans, whole-grained breads, pasta) and fruit (fresh, dried, or canned).

- Maintaining the usual level of fat, emphasizing healthy fats in nuts, seeds, and oil while minimizing saturated and hydrogenated fats.

- Consuming high-carbohydrate energy foods during and immediately after work.

Carbohydrates are the key, because they serve as the main energy supply for the body and are most easily utilized as a source of fuel. Besides their benefits as an energy source, as a countermeasure for fatigue, and for the immune system, carbohydrates have also been shown to maintain mental performance during prolonged periods of physical stress (Puchkoff et al, 1998). They are brain candy, man! The low carb diets that have been the fad recently will not meet the nutritional requirements for most firefighters, and should be avoided, at least during the fire season.

Apart from full, balanced, regular meals, Dr. Sharkey recommends high-carbohydrate energy foods or supplements during and immediately after heavy work so as to maintain and build carbohydrate stores in the body. In fact, the Missoula Technology and Development Center recommends that firefighters eat at least two energy bars daily: one between lunch and breakfast, and the other between lunch and dinner. Research has not indicated the superiority of any particular brand of energy bar. Clif Bar, Gatorade Bar, Harvest Bar, and Power Bar all meet the specifications recommended by MTDC.

Some candy bars contain an adequate supply of carbohydrate, and can be utilized as a source of energy. The downside is that they tend to be high in saturated fat, so their usage should be limited. The other problem is that since most of them are made out of chocolate they tend to melt. Go figure. Hot day, hot fire, Snickers soup. Who would have thought? That is why so often in our sack lunches we end up with either Salted Nut Rolls or Skittles.

If candy bars are to be used to provide supplemental carbohydrates, it is recommended that they have at least 30 grams of carbohydrate per serving. Again, a simple check of the nutritional label on an item will provide this information. Skittles, for instance, provide about 54 grams of carbohydrate per the standard 2.17 ounce package. That is about 17% of the recommended daily allowance for carbs. Talk about your rocket fuel! However, there is a catch. Included within these 54 grams of carbohydrates are 43 grams of sugar per package, which leads to about 230 calories for this same one cup serving. Nut rolls fare a little better. The 1.8-ounce bar that so often ends up in your sack lunch has about 27 grams of carbohydrates, and only 240 calories. That's why three energy bars should be eaten for every one candy bar, when candy bars are to be used to help meet energy requirements.

So with all these percentages, figures, and facts floating around, just what can a firefighter do to meet his or her nutritional needs without having to bring a calculator to every meal or snack? Here are some general suggestions:

- During fire season, or whenever engaged in arduous endeavors, increase your intake of complex carbohydrates (potatoes, corn, rice, beans, and whole-grain breads and pasta) and fruit (fresh, dried, or canned). In general, look for products that use whole grains. Put these on your plate first, and it will leave less room for the fatty stuff or the foods with only simple carbohydrates.

- Try to eat four servings of fruit (a serving is about the size of your fist) and five servings of vegetables daily (two and one-half cups). This will be challenging when you are relying upon others to provide your meals. It might be necessary for you or your crew supervisor to pack your own fruits and vegetables (fresh, dried, or canned) when you are engaged in fire operations.

- Utilize energy bars between meals to obtain needed carbohydrates. Candy bars can be eaten but should be limited because they tend to have high levels of fat and simple sugars. Because recent studies have shown that the most effective nutrition is to eat small quantities frequently during the work shift, the U. S. Forest Service is currently experimenting with changing the traditional brown bag lunch to "shift food" that contains a number of smaller food units to be eaten throughout the day. In other words, rather than eating your lunch in one sitting, try snacking on it all day long.

- Consume 300-500 kilocalories of carbohydrate energy during the two hours immediately after the work shift. Studies show that the replacement of carbohydrates is most rapid and effective within this two-hour window. Energy can come in a solid or liquid form (liquid carbohydrates will be covered in the hydration section). Some form of energy bar or drink should be consumed if more than two hours will pass after heavy work before the firefighter can eat a complete meal (due to transportation, other assignments, and so forth).

- Get in the habit of carrying your own food sources, be they energy bars, dried or canned fruits and vegetables, whole grain products, and so forth. Then you will not be as reliant upon others to obtain the fuel that your body needs to do this difficult and demanding job.

Hydration

If we use the analogy that food serves as fuel for the body, then liquids are a kind of cooling system. As most firefighters can attest, we drink huge amounts of water when sweltering on a hotline on a hundred-degree day. When one works in these kinds of conditions, adequate hydration can mean the difference between life and death. We all know we must stay hydrated. But questions still persist, such as What to drink? How much to drink? and How often?

As we perform our duties as firefighters—whether in the deserts of the Great Basin, the Northern Rocky Mountains, or elsewhere—our body sends blood to the skin, where over five million sweat glands produce sweat. As sweat evaporates, it helps to cool the body. Perspiration is simply the outer expression of the body's internal temperature regulation process. Through the sweating process, the firefighter's body loses a significant amount of fluid, oftentimes at a rate of more than one liter per hour for some individuals. (For reference, a liter contains about 35 ounces, or just slightly more than a 32 ounce quart.) If sufficient fluids are not replaced in the body, this temperature-regulation process can begin to fail. Heart and circulatory functions become impaired, and heat disorders (heat cramps, heat exhaustion, and heat stroke) may result.

- Heat cramps, the least serious, are involuntary muscle contractions caused by failure to replace fluids or electrolytes. Heat cramps usually affect muscles fatigued by heavy work, such as calves, thighs, and shoulders. They typically occur during exercise or work in a hot environment, or they may begin a few hours after a heavy work period. Heat cramps seem to be connected to heat, dehydration, and poor conditioning, as opposed to lack of salt or other mineral imbalances, and they usually go away with rest, drinking water, and a cool environment.

- Heat exhaustion, also caused by heat and inadequate fluid intake, is characterized by paleness, weakness, extreme fatigue, nausea, headaches, and wet clammy skin. Rest and water may help in mild heat exhaustion, and ice packs and a cooler environment (with a breeze or fan, if available) may also help. More severely exhausted patients may need IV fluids, especially if vomiting keeps them from drinking enough.

- Heat stroke, the most serious of the three, is considered a medical emergency due to failure of the body's heat regulating mechanism. The sweating process usually stops, and body temperature rises dramatically (106 degrees Fahrenheit or higher). Heat stroke is typically characterized by hot dry skin, and symptoms may include mental confusion, loss of consciousness, convulsions, and, in the worst case, coma. Individuals with heat stroke need to have their body temperature reduced quickly, often with ice packs or very cold water, and must be given IV fluids for rehydration. Fanning the victim will help promote evaporation, as will partial submersion in cold water. Persons experiencing heat stroke must receive advanced medical attention as quickly as possible (emergency medical services are appropriate here); and they may need to be hospitalized for observation since many different body organs can fail in heat stroke.

When heat stress conditions exist (which can be a vast majority of our time as firefighters), workers must pay close attention to the way they work and exercise, and make modifications as necessary. Sharkey recommends that, when possible, firefighters do such things as:

- pace themselves

- avoid working too close to heat sources for prolonged periods of time

- accomplish harder work during cooler morning and evening hours

- change tools or tasks to minimize fatigue, and

- take frequent rest breaks during work

Through these kinds of efforts, the impact of a highly heated environment can be better mitigated.

Of course, the single most important thing firefighters can do for their body is to pay close attention to its hydration needs. One rule of thumb for firefighters is to consume one liter of fluid for each hour of vigorous exertion in high temperature environments. This equals more than four gallons of water over the course of a sixteen hour shift! Granted, we are not always working intensely for sixteen hours straight, so this one liter per hour guideline probably will not hold true for an entire shift. Additionally, sweat rates are highly individualized, and some firefighters will require much more water than others. But, we still need to drink a large quantity of liquid each work-day in order to meet the body's hydration demands.

Sports drinks, such as Gatorade, PowerAde, or All Sport, can be used to meet some of these fluid needs. While similar to water, they also replace such things as carbohydrates, sodium, and potassium, all of which are lost or consumed as the body functions in the work environment. MTDC recommends that sports drinks comprise one-third to one-half of a firefighter's fluid needs. This percentage can be fulfilled through pre-packaged drinks or powdered drink mixes which allow firefighters to concoct their own mixtures to suit their individual tastes.

While MTDC recommends sports drinks for up to half of total hydration inputs, the optimal ratio remains an active area of research. The ideal percentage of water- to-sports beverages will undoubtedly vary from person to person. In the end, it becomes the responsibility of each firefighter to pay attention to his or her own body, and decide what hydration inputs best meets his or her individual needs.

An important generality to remember is that thirst almost always underestimates actual fluid need. Put another way, firefighters should be drinking much more often than when their body tells them to. Before starting a work shift, firefighters should be consuming water, sports drinks, or juice in quantities from a few cups to a quart or more. (At this point, research does not support hyperhydration — the practice of drinking large quantities of liquid in one sitting in an attempt to temporarily store extra fluids). During the shift, firefighters should be taking frequent fluid breaks. In the first two hours after work, in order to rehydrate the body and to replace carbohydrate stores in the liver and muscle, firefighters should consume about two quarts of a sports drink.

Staying adequately hydrated remains one of the most important physical challenges that we face as firefighters. Incorporating the following suggestions should help you to meet your body's hydration demands:

- Remember that thirst almost always underestimates your fluid needs. That's why firefighters need to take frequent fluid breaks, and probably drink more fluid than they think they need to. In the first two hours after a shift, firefighters should drink about two quarts of liquid (preferably a sports drink) to help the body rehydrate. However, while dehydration remains the main worry for firefighters, overdrinking (hyperhydration), especially drinking too much water without electrolytes, can cause problems. Seldom is more than one quart an hour required for long periods.

- Sport drinks should comprise part of each firefighter's hydration inputs, since they contain carbohydrates, sodium, and potassium, all of which are lost through sweat or burned through vigorous work or exercise. Firefighters will benefit if they get in the habit of carrying powdered sports drink mixes, which allow them to make their own concoctions.

- Crew supervisors and logistics personnel need to be aware of the hydration demands of their personnel, and provide adequate supplies of both water and sports drinks to those working under them. That being said, individual firefighters need to carry enough water with them to meet their own hydration needs for each work shift. Camelbaks™ or other personal hydration systems have been shown beneficial, both in terms of frequency of drinking behavior and total volume consumed, so their use should be carefully considered.

- A person's level of hydration can be assessed by observing the color, volume, and concentration of urine. Dark yellow-to-brown-colored urine, or low total volume of urine, indicate dehydration. Peeing into a white cup can help to assess the color accurately. Recent research from MTDC suggests that best way to assess hydration is to weight yourself pre-shift and again post-shift. Loss of greater than 2% body weight is considered dehydration and you need to drink more. Each morning, your weight should be back to what it was the prior morning.

- Firefighters need to constantly monitor themselves and co-workers for heat-related disorders, and when distressed follow the recommended mitigating steps.

Ergogenic Aids

Ergogenic aids are practices, devices, or substances that enhance a person's capacity for energy use, production, or recovery. Under such a broad definition, ergogenic aids can come in many different forms. They can be of a mechanical nature, such as heart rate monitors, weights, or nasal strips. Psychological techniques, such as imagery, music, hypnosis, or relaxation strategies, are also considered to be ergogenic aids. Acupuncture, massage, and whirlpools fall under the heading of physiological ergogenic aids. But perhaps the most commonly recognized form of ergogenic aids are pharmacological agents, ranging from the common to the complex – from caffeine and creatine to steroids, human growth hormone, EPO, and the broader classification referred to simply as supplements. This section will focus specifically upon pharmacological items.

Pharmacological ergogenic aids generate much discussion and controversy. Sports sections of every newspaper have carried many stories on the use and abuse of performance-enhancing substances. No sport is immune, be it the steroid issues surrounding major league baseball, professional football, and track and field, or the current EPO controversy within the world of professional cycling. Of course, many ergogenic aids of a pharmacological nature are legal and available over the counter. So, the more we know about them and their effects, the better off a firefighter will be should he or she decide to use them.

Caffeine

Caffeine is a drug that works as a mild stimulant to the central nervous system. While it is not believed to be addictive in a physical sense, it can be habit-forming from a psychological standpoint. No other substance is as synonymous with firefighting as good ol' caffeine. While it comes packed in various soda drinks, by far the most common form of caffeine is in the old standby, the cup of coffee. Simple to pack around and brew, coffee is as common in a firecamp or on the fireline as a Pulaski or a pair of White boots. As Edward Abbey once said, "Our culture runs on coffee and gasoline, the first often tasting like the second." Despite its popularity, though, there continues to be considerable debate about the impact coffee has on the user.

In its June 2004 issue, *Wildland Firefighter* magazine included an article about caffeinated beverages. The article argued that firefighters should avoid caffeine due to its diuretic qualities (diuretics comprise a large group of substances or drugs that act upon the kidneys and result in increased urinary output). Well, for many, such a statement is nearly akin to yelling "shark" at a crowded beach. Not surprisingly, the anti-caffeine position has been criticized by some in the fire community. Chuck Sheley, managing editor of *Smokejumper Magazine*, states in the April 2005 issue:

> *The bottom line is this: Caffeine is proven to sustain and even improve mental and physical performance during extended periods in which getting the proper amount of sleep and rest is not realistic — the kind of conditions firefighters face all the time. Certainly the increase in a firefighter's mental prowess from caffeine ingestion would lessen the potential for accidents. Moreover, evidence of a negative diuretic effect is suspect and minimal and could even be negated by supplementing caffeine intake with extra water. (p. 6)*

As noted by Sheley, there is no research evidence which supports the notion that caffeine in a beverage form is dehydrating. While caffeine does possess diuretic qualities, the effects are usually compensated for by the beverage's fluid content. As with most things, moderation is the key with regard to caffeine. Most experts agree that 300 milligrams of caffeine (about the amount found in three cups of coffee) is a moderate intake. Consumption levels of 600 milligrams or more have been associated with such things as tics and muscle tremors, headaches, heartburn, anxiety, restlessness, depression, and sleep-pattern disruption. So, if a firefighter should find himself or herself ingesting caffeine in more than moderate amounts for extended periods of time, he or she should attempt to cut back intake to a more moderate level.

Nicotine

As everyone knows, nicotine is the primary drug within any tobacco product. Like caffeine, nicotine works as a central nervous system stimulant, with effects such as increased mental alertness and reduced levels of fatigue and drowsiness. This sounds great, but as with most substances, there is a downside. Nicotine is highly addictive, and tobacco usage in general is strongly associated with several types of cancer. While this author knows of no studies which have actually looked into the prevalence of tobacco usage amongst our profession, one need only hang around firefighters to realize that it is an extremely popular drug within the ranks, be it cigarettes or the much more common form of smokeless tobacco.

As an on-again, off-again chewer for years, it would be hypocritical for this author to denounce the use of tobacco. As a firefighter, working outdoors and pulling long shifts on the fireline, I have found chewing to be an immensely pleasurable activity. I know that it's bad for me, and that I should probably stop doing it. I have quit several times (usually during the winter), but when fire season rolls around, I inevitably end up buying that first can of chewing tobacco at the convenience store, and the race is on. To me, the use of tobacco is a highly personal choice. If a person decides to do it, that is his or her right.

With all this in mind, what follows are some suggestions regarding the use of tobacco products:

- As with most everything else in this life, moderation is the key. If you choose to be a tobacco user, keep it reasonable. This will be extremely difficult, though, because nicotine is highly addictive. Try to limit yourself to a set number of chews or smokes per day, and keep this number as low as possible. Obviously three packs of cigarettes a day, or a snoose in your lip 24/7 is not being moderate.

- Nicotine can have a dehydrating effect on the body, so if you are chewing, realize that you will need to be drinking more liquids to stay hydrated.

- If you use tobacco, be responsible. Do not leave your butts lying around, and don't leave half-full bottles of chew spit for somebody else to pick up. Avoid throwing that large wad of chaw in the drinking fountain or some other conspicuous spot. If you smoke, be considerate when doing it around those who don't smoke.

- Try not to be a moocher. Nobody likes a chronic mooch! If you use it, then buy your own. If you bum an occasional pinch or smoke off somebody, that's okay, but then buy them a can or pack to re-pay the favor.

Supplements

Set foot in just about any grocery store, pharmacy, or natural foods store, and you will find them. Open your email and chances are good that you will have been spammed by somebody trying to sell them. Visit a specialty store, like GNC, and you will be overwhelmed by the sheer number and the variability of products available. Of course, the goods being referred to are none other than supplements. While they can come in herbal, vitamin, or mineral varieties, the Food and Drug Administration (FDA) places them all within the category of "dietary supplements."

Although no data is available regarding the use of supplements within the wildland firefighter community, a recent study of the U.S. population as a whole found that nearly 20% were currently taking some form of dietary supplement (Kelly et al, 2005). It is reasonable to assume that a similar percentage of firefighters are purchasing and using these products. For those doing so, however, there tends to be a great deal of misinformation regarding their benefits and little information about their possible negative effects.

The FDA is the governmental agency responsible for overseeing the safety of U. S. food and drug products. In 1994, Congress passed the Dietary Supplement Health and Education Act (DSHEA), which limited the FDA's control over those products labeled as "dietary supplements." With the DSHEA in place, manufacturers do not have to prove the safety or efficacy of a product before placing it on the market. Through the passage of this act, Congress placed greater responsibility upon the consumer to decide what, if any, products to use. Consequently, dietary supplements are not subject to the same rigorous standards the FDA uses to evaluate prescription drugs or those sold over the counter. Supplements can be marketed and sold with limited proof of their effectiveness or safety, and vendors can make health claims about their products based upon their own reviews or interpretations of research, all without FDA authorization. That being said, the FDA does have the ability to take a product off the market if it's proven to be dangerous, as it did with Ephedra.

The most common dietary supplements are the multivitamins/multiminerals, such as Centrum®, Puritan's Pride®, and One-a-Day®, and a host of other brand names and generics. Since as firefighters it can be challenging at times to meet our nutritional requirements, some might be tempted to rely upon supplements to make up the difference. However, most major organizations that have investigated the role of vitamin and mineral supplements (American Dietetic Association, American Heart Association, American Cancer Society, American Academy of Family Physicians, Mayo Clinic) have advocated for a balanced and nutritious diet based on fruits, vegetables, and grain instead of reliance upon supplements.

However, as noted in the Spring 2002 *Wildland Firefighter Health and Safety Report*, supplements may be required by athletes who restrict energy intake, who use severe weight-loss practices, who eliminate one or more food groups from their diet, or who consume high carbohydrate diets with low micronutrient density. As firefighters we may find ourselves meeting some of these criteria. Vitamin and mineral supplementation therefore becomes an individual choice. But at this point in time, there is no solid research evidence supporting the usage of nutritional supplements for wildland firefighters. Should you decide to use them, it should be done with caution, and only after a careful evaluation of the product's safety, efficacy, and potency. If you do use a supplement, megadoses of vitamins and minerals are not recommended due to the possibility of toxicity and adverse interactions among nutrients.

Creatine monohydrate, or simply creatine, has become one of the most popular supplements on the market today. It is naturally synthesized in the body from amino acids found primarily in the liver and kidneys. Using the circulatory system for transport, it moves through the bloodstream and is then stored in muscles. Creatine is most commonly found in fish and meats, and most adults consume about 1-2 grams of it daily in their normal diet. In the 1990s, creatine gained immense popularity as a "natural" way to build lean body mass and improve overall athletic performance, and that popularity continues.

Again, no data exists as to the prevalence of creatine usage within firefighter ranks. Current published reports indicate that upwards of 25% of professional baseball players and 50% of professional football players take some form of creatine supplement. It also remains extremely popular amongst adolescent athletes, many of whom have admitted taking doses far higher than recommended levels. With usage numbers this high, then, one can presume that at least a portion of the wildland fire community is utilizing this supplement.

Even though it has become popular, research on creatine has failed to substantiate either benefits or drawbacks of usage. More than 300 studies have evaluated the ergogenic value of creatine supplementation, and about 70% report statistically significant results from usage. Although this percentage of positive results is high, methodological problems with many of the studies make it difficult to form firm conclusions at this time.

Several studies have suggested that creatine may improve muscle mass and strength in both men and women, especially when its usage is done in conjunction with increased physical activity. It has also been suggested that creatine may enhance athletic performance by increasing the time it takes for muscles to fatigue. While many researchers believe that creatine may be useful for short-duration, high-intensity exercise, it has not demonstrated effectiveness for endurance sports (those that require sustained aerobic activity). Since wildland firefighting requires endurance, the effectiveness of creatine usage by firefighters remains in question.

That being said, with 70% of all studies demonstrating significant positive results from creatine supplementation (even though there are methodological concerns with much of the research), it does appear there may be some benefits to taking creatine. It boils down to personal choice. If a firefighter believes that the cost is worth the potential benefits, it remains his or her option to use it. Once again, moderation is the key. The majority of studies investigating creatine usage have incorporated daily dosages of between 10-20 grams. Therefore, usage at levels above these amounts is strongly cautioned against. With so much ongoing research into creatine, it is also recommended that anyone deciding to use it pay careful attention to the results of future studies.

With many dietary supplements currently available, it is beyond the scope of this performance guide to report on each and every one. If a firefighter is interested in supplementation, what follows are some general suggestions regarding their usage:

- If you are taking any other prescription or over-the-counter (OTC) medications, talk with your doctor about the use of any supplement. Adverse drug interactions can occur, with possible serious side effects.

- If you are pregnant or nursing, the rule of thumb is not to take any medication (prescription, OTC, or dietary supplement) before consulting with your doctor.

- For those who must undergo surgery, certain supplements can interfere with the effectiveness of anesthesia, or cause dangerous complications (such as bleeding or high blood pressure). Again, consult with your doctor prior to surgery.

- Know what you are buying. Look for the U. S. Pharmacopoeia's "USP Dietary Supplement Verified" seal on the product's label. This indicates that it has met certain manufacturing standards for uniformity, cleanliness, and freedom from environmental contaminants. Do not assume that anything you buy is safe.

- Use only those products which have been scientifically tested. If it has, it will usually say so on the label. If you are still unsure, a simple Google search of the product should provide additional information regarding usage of that particular supplement.

- Beware of product claims that are too good to be true. Very few things in this life are.

The job of wildland fire suppression remains a difficult one. It can be and is extremely taxing upon our bodies. Therefore, the better care we take of our "temple" the more able it will be to respond positively to the demands we place on it. Through proper nutrition and hydration, and the moderate use of ergogenic aids (if one opts to utilize them), the wildland firefighter can optimize his or her physical functioning, and be that much closer to the goal of overall optimal performance.

5

Injury

August 1, 2001, 1900 MST: Lying on a rocky slope north of Yellowstone National Park, it occurred to me that this latest jump might very well be my last as a smokejumper. I had to laugh at myself, despite the pain inside. The irony of the situation was almost comical. My main goal in coming back this year had been to make my 100th jump. Now, on my 99th, things had just gone terribly wrong.

The fire call came into the Missoula fire depot late in the afternoon. Sixteen of us loaded up on the DC-3 and headed east. The Hoppe Fire was fifty acres and running by the time the jumpship was overhead. Retardant planes were dropping their loads, so we racetracked above them, waiting for our chance to go in. Jumping third in a three person stick, I ended up quite a ways out from the jumpspot that had been chosen. The wind that had been blowing steadily from the west was now nothing more than an occasional puff. I had counted on this breeze to help push me back toward the clearing, but now I would not have its assistance. Trying to get back to the jumpspot, I decided I was going to go all-out to get there. The alternate landing zones were rockslides and stubby trees, none of which I wanted any part of. By the time I finally realized that I was not going to make it to the original jumpspot, my options for picking out a suitable alternate LZ had been greatly reduced. I made a late turn in order to avoid smacking directly into the hillside. While this positioned me parallel to the slope, the canopy did not have time to regain its forward momentum. I tracked sideways across the mountain at a high rate of speed. Coming down, I struck a downed dead log with the right side of my body, and flipped over it into a pile of large granite rocks.

Immediately I could tell that I had inflicted damage to various parts of my body. Worrying about internal injuries, I peeled off my jumpsuit and started limping up to the ridge where I knew that some of the other jumpers had landed. I needed to find some bro's. They would help me. They could fix me, let me know that everything was going to be okay. A trauma kit was thrown from the plane, full of medical supplies. Rick and Godot became my EMT's. They took my vital signs, taped frozen canteens to my throbbing leg, and told me jokes to take my mind off the pain. I could not help but think how my jump career had probably just ended. What resulted was a confusing mixture of shock, anger, and sadness over the events that had just transpired. I did not want my jump career to end this way. I wanted to leave smokejumping on my own terms. Now it appeared the decision to leave had just been made for me.

When the military rescue helicopter finally arrived about 0100 early the next morning (its pilots using night-vision goggles to navigate the mountainous terrain), my spirits soared. I was going to get off this mountain, and hopefully live to fight fire another day. I was flown to Belgrade, Montana, then transported via ground ambulance to the Bozeman emergency room. There, I learned my injuries were not as severe as I initially feared: I had a broken leg, some broken ribs, but no internal organ damage. The next morning, following a fitful night of sleep in an ER bed, I was loaded up on the DC-3 at the Bozeman airport and flown back to Missoula. My jumping was definitely over for the season. What awaited me was lots of limping, a healthy dose of physical therapy, and "flying a desk" duties for the remainder of fire season. Fortunately, I did recover, though, and eventually earned my 100-Jump pin early the next season.

Wildland fire suppression occurs in a high-risk environment. Fires, by their very nature, can be dangerous and unpredictable. On the line, we are exposed to such things as erratic fire behavior, rough terrain, falling snags, and sharp tools. Each of these has the ability to hurt or even kill us as a worst-case scenario. However, we are not just at risk on the fireline. We are also exposed to hazards and possible injury when doing any of the other duties which we are assigned, be it project work or even physical training. The list of possible injuries and ways to be injured is long.

Firefighters share many traits in common, one of them being a tendency toward an active and busy lifestyle. Be it skiing or snowboarding, mountain biking, rock climbing, kayaking, hunting, or running, as a group we are drawn to activities that challenge us physically. While such challenges help us to work and play simultaneously, they also carry risk. Strenuous physical activities of any kind open us up to the possibility of injury. At some point in our lives, most of us will incur an injury that is significant enough to curtail our normal routine. In other words, firefighters come in two models: those who have been injured, and those who have not been injured yet.

Of course, the career impact of any injury is directly related to its severity. For example, a sprained ankle will probably not require any specialized medical assistance. With time and treatment, the swelling in the ankle will eventually subside enough to allow greater range of motion. Typically, in a few days the injured person will be able to resume some daily activities, sometimes at a much slower rate. As the injury heals, the injured person may return in baby steps to activities he or she was doing prior to the injury.

Conversely, a fractured femur will necessitate a great deal of medical intervention. Surgery may be in order to pin or screw the bone back together. Physical therapy may be needed to re-build lost muscle mass and to maintain range of motion in other parts of the body. A long period of rehabilitation may be needed before the body is fully healed.

The person with a sprained ankle may only miss a few days of work, if any. However, the person with the fractured femur could be out for six weeks or more, maybe several months. If the fracture is severe enough, the injured person may be faced with the prospect of never again attaining the level of mobility which he or she was at prior to the injury. Clearly, some injuries create more havoc on the body and lifestyle of the firefighter than others.

Aside from physical effects and time spent in rehab, injuries can impact the psychological functioning of injured persons. Routines that were normal now become disrupted. Activities like washing or dressing that were previously done quickly and without thinking are now not possible, or possible only in a limited way and with pain as a constant reminder. Psychologically, injuries may lead to such things as self-doubt and a loss of confidence. Anger, depression, confusion, and apathy are some other possible psychological impacts of injury.

Environmental factors also might come into play when a person has been injured. Support systems (friends, family, team members) that were in place before the injury may not be available, either because they do not know how to help or they simply choose not to. The stabilizing effect of being a member of a team may vanish as the injured person is left to recover on his or her own while the rest of the team continues with the fire season or with regular work activities.

So the injured person must not only cope with the primary physical impact of the injury, but also secondary psychological and environmental effects as well. These secondary effects may impact the efficacy and speed of physical recovery. Injury cuts across categories, as it may impact all three factors of optimal performance – physical, psychological, and environmental. Although this chapter is included within the physical factors section, it could just as well be placed in the psychological factors section or the environmental factors section.

Since firefighters share many traits in common with athletes, information from athlete-oriented disciplines of exercise physiology and sport psychology can help when it comes to firefighter injury.

Treating Minor Injuries

With any injury, physical treatment is the first priority. If you are injured, the most urgent priority is to stop what you are doing so as to prevent further damage. Most people instinctively know the difference between pain and injury. If you think you are hurt, you probably are. Check the "tough guy" and "tough gal" image at the door. Take a minute and evaluate the hurt.

For the treatment of and recovery from minor injuries, the acronym **RICE** is familiar to many. The acronym stands for **Rest, Ice, Compression,** and **Elevation.**

Rest is just what the name implies, and ties back in with stopping what you are doing if you think you've just hurt yourself. Rest also can come in the form of a splint, cast, or crutches to immobilize an injured part of the body. And, rest also refers to getting adequate sleep, so that the body has sufficient energy needed to repair itself. Rest does not mean lying on the couch for two weeks after you have slightly twisted your ankle. Most joint injuries will need to have some movement in order to stimulate healing and growth of new tissue.

Ice is often the second step in the treatment process. Icing helps reduce swelling and pain from an injury. The application of ice is most effective when done immediately after sustaining an injury. Effectiveness of icing is greatly reduced 48 hours after most injuries happen, so get the ice on it quickly. When cold is applied to an injured area, it deeply penetrates the soft tissue. This slows down the blood flow to the area, which reduces swelling and causes a numbing of the nerve endings surrounding the injured region.

Icing results in a predictable sequence of sensations: an initial cold feeling, followed by stinging, then burning, and finally, numbness of the iced area. For many people, icing can be uncomfortable; but, to achieve any benefit, it is important to endure the cold. Fifteen to twenty minutes is the maximum amount of time that an injury should be iced per session. Each icing should be followed by a twenty minute "warming" period in which the ice is removed. When treating mild injuries, the "ice on/ice off" process should be repeated for two-hour periods, twice a day, for the first 72 hours. The icing process may be longer for injuries that are more severe.

Ice may be applied in several different ways. A bucket filled with a combination of water and ice is particularly effective, especially for ankle or wrist injuries. Plastic baggies of crushed ice or even sacks of frozen vegetables can also be used. Ice massages, done by gently rubbing the injured area with ice, are particularly helpful for reducing the pain and swelling. Care must be taken to avoid "frostbiting" the affected area. A thin towel or piece of cloth can be used between the injury and the ice, and the twenty-minute time limit should be strictly followed.

Compression, the third component of the RICE treatment program, also helps to reduce the swelling that accompanies most injuries. An Ace bandage or other type of elastic wrap can be used to accomplish this. When wrapping, begin at a point below the injury, and wrap upward. In other words, start the wrap from the furthest part of the body and proceed toward the closest. For example, with an injured knee, the wrap should initiate near the upper calf region, and finish at the thigh. If heavy throbbing results, simply re-apply the wrap, but in a looser fashion. Wraps can be worn during periods of activity, just remember to ice and elevate the injury before and afterwards. Do not wear the wrap while sleeping.

Elevation is the fourth element of RICE. Elevation helps to reduce the pain, swelling, and bruising that occur by draining the fluids that accumulate in the injured area. If possible, apply ice while elevating. Keep the injury above the level of the heart, as this also helps inhibit swelling.

Maintaining mobility of the injured area is part of RICE treatment and recovery. Keep in mind, however, that returning to full activity prematurely may slow healing and lead to re-injury. Since you know your body, you have to be the judge. Mild discomfort during activity is not uncommon after minor injury, and under a doctor's supervision should not pose too much concern. However, moderate to severe pain, or pain that persists for a long period of time, is a signal that your body is probably not ready for that level of movement. Slow down, and allow your body more time to heal.

For many injuries, improvement begins to occur within the first 48 hours of RICE treatment. Remember that healing time varies from person to person, and also depends upon the severity of the injury. If pain or swelling does not decrease within the first couple of days, see your primary care physician or go to the emergency room, depending on the severity of the symptoms. Once injuries do begin to heal, heat can be applied to increase blood flow into the damaged area, which promotes the healing process.

Injuries that are more severe will require specialized medical assistance. To deal with the physical impacts of this damage to the body, it is important to seek out and obtain help as quickly as possible. With professional medical advice, proper diagnosis, treatment, and rehabilitation (if necessary) can be undertaken. Again, for any injury, dealing with the physical component is the first priority.

The Psychology of Injury

For many years psychologists have studied the impact of injury on the overall functioning of athletes. Richard Suinn (1967) was one of the first psychologists to suggest that athletes cope with injuries in basically the same way that any person faced with physical impairment or significant loss does. Suinn proposed a chain of possible reactions to the injury, including initial shock, denial of the injury or its severity, depression or anxiety once the injured person understands the severity/impact of the damage, and finally partial or complete acceptance of the injury.

Other sport psychologists have elaborated upon Suinn's work, and have suggested that athletes experience a grieving process when an injury leads to temporary or permanent loss of their athlete identity (Rotella, 1984; Astle, 1986; Rotella & Heyman, 1986). This grieving process follows the stages outlined by Elizabeth Kübler-Ross in her best-selling book from 1969, On Death and Dying. From her work with dying patients and their families, Kübler-Ross identified a series of predictable stages that people move through, including denial of the death, anger, bargaining, depression, and lastly acceptance of the death as a permanent condition.

Subsequent research by others has suggested that stage models (like Kübler-Ross) may not accurately represent the path an athlete takes in response to injuries that result in temporary physical impairment. Recent research has suggested that athletes following an injury usually experience a brief period of mood disturbance (marked by increased tension, depression, anger, and decreased levels of vigor) but that they return to normal when they believe that they are on the road to recovery (McDonald & Hardy, 1990; Smith, Scott, O'Fallon, & Young, 1990). Grief reactions similar to those outlined by Kübler-Ross may very well occur for those athletes who have suffered career-ending injuries, but research does not firmly support this notion at this time.

Other research with athletes has shown that feelings of loneliness and separation may occur after an injury (Crossman & Jamieson, 1985; Lewis-Griffith, 1982). Athletic teams do many things for their members, and one of the most important is a social support system. If an injury curtails a person's ability to practice or participate with the team, the support network can be disrupted. Injured individuals may begin to feel that they are no longer a contributing member. They may disengage from team activities, even though this is a time when an injured person needs the support. Likewise, the non-injured team members may pull back, not knowing what to say or do for the injured person.

Injuries lead to new stresses and anxiety. The injured person may be faced with nagging questions: Why did this happen to me? Will I recover? Will I regain my place on the team? If and when I return, will I possibly re-injure myself in the same way? What am I going to do if I do not heal properly? How am I going to make up for the money this injury is costing me? Since the answers to these questions are often uncertain, anxiety can result. With many unknowns piling up, the injured person may begin to think about "worst case" scenarios. With the mounting uncertainty, injured individuals may attempt to come back too soon before the injury has had a chance to fully heal. Performance may suffer, either because he or she cannot meet the physical demands of the task at hand, or because he or she is so worried about re-injury that overall situation awareness decreases. Both are recipes for bad things to happen.

Post-Injury Warning Signs

Just as we have tactical "Watch Out" situations for the fireline, a similar type of list can be used regarding injury. Sport psychologists Petipas & Danish (1995) have come up with a list of warning signs that may be helpful in identifying problematic post-injury adjustments for the injured athlete. This list has been modified so that it applies more directly to wildland firefighters.

- Evidence of anger, depression, confusion, or apathy

- Obsession with the question, "When will I be able to fight fire again?"

- Denial, reflected in remarks such as, "Things are going great," "The injury is no big deal," or other comments that lead the listener to believe that the firefighter is making an extraordinary effort to convince you that the injury does not really matter.

- A history of coming back too fast from injuries.

- Exaggerated storytelling or bragging about accomplishments either on or off the fireline.

- Dwelling on minor body complaints.

- Remarks about letting other team members down, or feeling guilty about not being able to contribute.

- Withdrawal from teammates, other firefighters, supervisors, friends, or family.

- Rapid mood swings or striking changes in affect or behavior.

- Statements that indicate a feeling of helplessness to impact recovery.

If you have been injured, and find yourself thinking or behaving in one or more of these ways, it might suggest difficulty in coping. The same applies if you observe a teammate displaying these attitudes. Realize that such thoughts and behaviors are entirely normal for someone who has suffered an injury.

Tips on Dealing with Injury

Injury can and does have an impact upon all three factors of performance — physical, psychological, and environmental. Therefore, when dealing with injury, each of these areas might have to be addressed during recovery. What follows are some general suggestions on how to handle an injury should it occur.

- Follow the Emergency Medical Technician (EMT) mantra, "Scene safety, scene safety, scene safety!" If you've been hurt, make sure that the area is safe enough that you or someone else will not be at risk for further injury.

- Incorporate the RICE treatment program as the first step in recovery for most minor injuries. It will help to reduce the pain and swelling that results, thus speeding up the healing process.

- Another "no-brainer," but if you have been significantly injured, immediately seek out professional medical advice.

- If the injury has happened at work, be it on the fireline or elsewhere, get the appropriate paperwork filled out (CA-1, CA-2, witness statements, and so forth). If you work for the government, you know how important this step is. Talk with your personnel officer or the administrative person who deals with injuries. Even if you think it's only a minor injury, get the paperwork filled out. Oftentimes what we might think is not a significant injury can turn out to be much more serious than we originally believed it to be. Firefighters tend to be "pull-yourself-up-by-your-own-bootstraps" sort of people. We have a tendency to minimize pain and to try to keep personal issues to ourselves. This is not a good pattern in dealing with injuries. If you think you need help, get it.

- Get copies of everything associated with your injury, such as government forms, medical evaluations, leave slips, doctor's reports, and so forth. Create a folder for these materials, and keep it somewhere safe and where you can access it quickly.

- Utilize your support structure to help you. This includes family, friends, team members, and anyone else who can be of assistance in this challenging time. Try not to pull away from those who can help you. Stay involved with the team, no matter how tough this is to do. As firefighters, we are often closer to our teammates than we are to even our own families. For months on end we may travel, eat, sleep, and play together. Utilize co-workers to help in your recovery. If the injury has occurred to a team member, do what you can to "be there" for this person. Keep the injured teammate up to date on what is happening with the team. Phone them. Let them know that you are thinking of them and are looking forward to when they return to work.

- Keep in mind the stages-model of injury recovery. You might find yourself being angry, or depressed, or in denial about the injury. That is entirely normal. The key is to realize that recovery can be a long process. Strive toward acceptance of the injury. Once you are at the stage of acceptance, you can begin to work toward what you need to do for the next stage, which is recovery.

- Set attainable goals during your rehabilitation (for more information refer to the goals chapter in this book). Goals that emphasize effort and steady improvement will be better than goals that set a specific time-frame for their achievement. If the doctor tells you that your injury will take 4-6 weeks to heal, then 4-6 weeks is probably how long it will take. Doctors are experts in this. Resist the temptation to be a "hero" and come back in three weeks because you think you are sufficiently healed. By short-circuiting the recovery you greatly increase the risk of possible re-injury.

- Commit to your rehabilitation program. If a doctor or physical therapist has developed a rehab program for you, then "own" it, and stick to it. Although this might seem like a no-brainer, research shows that compliance (adherence) rates in therapy range between 40 percent and 91 percent. In other words, in some cases only 4 out of 10 injured athletes followed the rehabilitation program that was developed for them. Again, trust the professionals who deal with injured athletes on a daily basis. Doctor knows best (or was it Father knows best? Never mind!)

- Be ready to answer questions from others about what happened, how the injury occurred, its severity, and so forth. People are curious. Even if you aren't comfortable talking about your injury, have some sort of "pat answer" ready that you can tell people. If you are comfortable talking about the injury, then do so. The quicker you can accept the fact that you are injured, the faster you will take the steps needed (physically, psychologically, and environmentally) to get healed.

- Try to look at the opportunities the injury might provide, instead of focusing on the negatives. The world of fire suppression has many moving parts. If you cannot go out on the line with the crew, maybe you can find another place where you can be of assistance. For example, you can take time to learn about other areas of the organization that you are not familiar with. The important thing is to find something which allows you to feel like you are making a positive contribution. After I broke my leg on my 99th fire jump in 2001, I finished out the fire season working in our operations center. Although I missed being on fires with my teammates, I was in a position to help the smokejumping program by taking part in the logistical support needed to make the base run more smoothly. And I also learned a great deal about what goes on behind the scenes. In the end, I became a more effective firefighter because I had a better understanding of how the system operates. Finally, if there is no opportunity to work elsewhere in the organization, an injured firefighter might be able to use the time off as an opportunity to focus on other aspects of his or her life that may have been put on the back burner due to the busy demands of being a wildland firefighter.

- Get educated about your injury. Learn everything that you can about it—its character, its etiology, its treatment, the rationale and the goals of rehabilitation, the risks and benefits of various treatment options, and probable outcomes and expectations. With the world at your fingertips from a simple Google search, there is no reason not to gather such vital information. Your doctor or care provider is another source of input, as are people who have had a similar injury. Information is power. The fewer the unknowns, the better for you. If you are not satisfied with your initial quality of care—including the quality of information you receive—then seek a second or third opinion from other physicians or specialists.

- Stay in shape, if at all possible. As firefighters, we invest a great deal of time and energy into our physical conditioning. Much of our confidence and self-assurance comes from knowing we are generally in pretty good condition. Even with an injury, we can usually continue to work out other parts of our body. Granted, whether you can continue to work out will depend on the magnitude of your injury or injuries. If they are severe enough, rest and recuperation may be your only choice. However, if you have recently broken your arm, there are still many things you can do to stay in shape, such as lifting weights with the lower body, running/jogging, using the Stairmaster, and so forth. If you have a lower-extremity injury, you can still focus conditioning/training efforts on the upper body. Even doing a little will usually lift your mood and self-confidence. Obviously, you need to check with your doctor or care provider regarding what you should and should not do during your recovery.

If you have tried the above tips and are continuing to have a difficult time dealing with the injury or the stress that comes with it, seek out a mental health professional. Employee Assistance Programs (EAPs) are free of charge to all federal employees. If you are a state, local, or contractor-type employee, find out what forms of assistance are available to you, if any. If you do not know, check with your personnel office or human resources department. Be an informed consumer. Utilize professionals, be they counselors or sport psychologists, who specialize in sport psychology or at least have a good background in helping athletes deal with injury.

Life Stress and Injury

Although it may seem overly simplistic, the best way to deal with an injury is to avoid becoming injured in the first place. Years of research by sport psychologists suggest that a combination of conditions put athletes at greater risk of injury:

- negative life stresses
- an increase in daily hassles
- previous injuries, and
- poor coping resources

How so? The combination of these four factors results in what psychologists call an elevated stress response. According to Brown (2005), an elevated stress response leads to such things as increased muscle tension, increased distractibility, and a narrowing of attention so that the athlete is not as aware of or responsive to critical events or cues.

Major life stresses include such things as the breakup of a relationship, the death of a loved one, loss of a job, or simply moving from one house or apartment to another. Daily hassles are minor problems and irritations that disrupt one's normal routine. Both major life stresses and daily hassles can increase the likelihood of injury. The probability of injury also increases if the person who has been previously injured attempts to come back to work before the injury is fully healed. Even if fully healed from the injury, the chance for re-injury increases if the person is not psychologically ready to return (either because the athlete tenses other muscles to protect the injury or by avoiding situations that threaten re-injury). Poor coping resources include such things as a lack of stress management skills, unhealthy life habits, and little or no social support.

Awareness is the key. If you are experiencing major life stresses, or even an increase in daily hassles, realize that you are exposed to an increased risk for injury. Likewise, if you are attempting to come back too soon from a previous injury, you are prone to re-injury. And lastly, if you have poor coping resources, you also run the risk of being injured. The good news is that you can improve on such things as stress management and relaxation (see Chapter Nine in this book). You can adopt more healthy life habits, by doing such things as eating well, drinking less alcohol, and getting more sleep. And you can work at developing and maintaining a solid social support network to help you through difficult times.

Injuries can affect firefighters in all three performance areas: physically, psychologically, and environmentally. So, the more we know about how injuries impact us and what we can do to deal with them, the better we are prepared should an injury occur. By addressing the physical effects upon the body through the use of effective treatment programs, such as RICE, swelling and pain can be reduced, thus leading to quicker healing. If physical therapy is needed, adhering to the rehabilitation program will help in the recovery process. Getting educated about the injury, setting goals for recovery, and looking for the opportunities the injury provides as opposed to focusing on the dangers will help deal with the psychological side of injury. And, by utilizing our support systems instead of pulling away in time of need, the injured person will be able better to manage the environmental impacts of injury.

PART TWO: PSYCHOLOGICAL FACTORS IN OPTIMAL PERFORMANCE

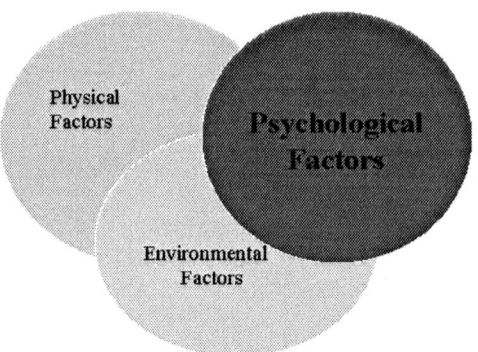

The deaths of the fourteen firefighters at South Canyon in 1994 had a tremendous effect upon the wildland firefighter community. While this event was a tragedy unmatched in our history, it has led to several positive changes, and for that we should all be eternally grateful for the ultimate sacrifice that these individuals paid. Their deaths have undoubtedly contributed to the saving of subsequent lives on the fireline, making them heroes in every sense of the word.

One of these positive impacts has been a focusing on the role that psychological and sociological factors play in wildland fire suppression. Beginning with the first Human Factors workshop held in 1995 in Missoula, Montana, the wildland fire community officially began to address the importance of these variables. From this "birth" of interest, we have seen several improvements: the development of leadership curricula and leadership-related resources for firefighters, improved communication capabilities on the fireground, and more effective training, to name just a few.

While interest has been growing for a decade, human factors have been an issue as long as humans have been organizing into groups to fight fires. Since optimal performance relies significantly upon psychological factors, the more we know about human factors, the more prepared we will be in the quest for the optimal performance zone. The list of human factors that come into play during fire suppression is lengthy, but this guide will focus on five in particular: Attitude, Goal Setting, Teams, Stress, and Transitions.

6

Attitude

July, 1994: "This assignment fucking sucks," Tool (name changed to protect the guilty) blurted out, a comment heard by most of the crew, but out of earshot of the Crew Boss and the other overhead. "We've been mopping up on this same division for three fucking days now! I haven't seen a smoke since this morning. What a fucking waste of time! We should be over on the Eastside putting in some hotline with those other crews. This is bullshit!"

For seven days now it has been like this. The Type II crew had formed a week ago, and from the beginning Tool has been a non-stop griper and complainer. The buses, the food, the overhead team, you name it, Tool bitches about it. Although clearly he has gotten onto just about everyone's nerves, nobody has yet said anything to him about his constant whining. Most of us on the crew have simply learned to tune him out. However, he has managed to exert some control over a couple of the younger and less experienced members of his squad. Now they are beginning to bitch about things nearly as much as Tool.

Seven days remained on the 14-day assignment, and for the majority of us on the crew, the next week cannot go by fast enough. I want to say something to the guy, but his attitude pisses me off so much that I instead have decided to ignore him. I know this is not the best approach. Still, I do nothing to challenge his attitude. When the assignment ends and we finally get back to Miles City, I vow that it will be a frosty day in Hellsville before I go out on a Type II crew again.

From a psychological factors standpoint, a positive attitude is one of the most important assets you bring to your job as a wildland firefighter. Let's face it: firefighting is a tough profession. With a positive attitude, we can handle adversity and challenges that confront us. There is an old saying: Attitude is everything. Although only three words, this axiom is powerful because it covers so much. Granted, a positive attitude alone will not guarantee optimal performance on the fireline. However, no matter how tough things get, with the help of a good attitude we can "grin and bear it." A positive attitude literally helps us over the hump of some of the worst hardships.

Unfortunately, the flip side of this saying also holds true. Negative attitudes do just the opposite of positive ones. Negative attitudes cloud our thinking, and put us into a reactive instead of a proactive mindset. Easy challenges become insurmountable blockades when our attitude goes sour. When we begin to focus on the negatives, they turn a harsh environment into a toxic waste dump.

If you have been involved in fire suppression for any length of time, you have probably been on a crew that had at least one person who was an extremely disruptive influence on the rest of his or her teammates. One person with a PPA (piss poor attitude) can blow a stinking cloud onto the whole team. People with negative attitudes can discourage those who work around them. Should there happen to be two or three individuals like this on a crew, their negative energy can become synergistic, with negativity feeding into more negativity. Inevitably, whether they want to or not, the rest of the crew can gets drawn into this downward spiral of pessimism.

The thing to remember about attitudes is that they are like herpes: highly infectious and contagious. What attitudes we have, others can easily catch. Think about your own experience working on fire crews, or other teams you have been on in the past. Undoubtedly, you can recall someone who just always seemed to be in a good mood (excluding the drug-induced). These kinds of people are usually fun to be around because they remain upbeat no matter what happens. Their positive nature rubs off on others.

Now, think about people like Tool, the bitcher and complainer we met earlier. Tool's negative attitude resulted in two less-than-desirable outcomes. At best, people tuned him out and stopped seeing him as a contributing member of the team. Theoretically, Tool might have had some knowledge or skill that could have helped the group, but because of his constant belly-aching people started to ignore him. So the team is hurt when one member's skills are not effectively utilized. But who can blame the rest of the team for blowing Tool off. I mean, who wants to deal with someone who bitches all the time?

The other possibility is that Tool gets others to buy into his negativity. And so in this way the "infection" spreads to other members of the team. Soon enough, Tool has poisoned others on the crew as they too have adopted his PPA. Now it becomes a non-stop bitchfest. For those who somehow manage to remain uninfected, the enjoyment of work completely evaporates because they are surrounded by constant complaining. Since in Tool's case we are talking about a Type II crew, the group will disband after their fourteen-day assignment. It can be a miserable two-week period, but at least it ends before long.

However, if we should happen to have a "Tool" on our shot crew, helitack crew, engine, jumplist, or IMT (Incident Management Team), we will have to deal with his or her negativity all season long (and maybe longer than that if his or her sorry ass doesn't get fired!) Since many crews spend a considerable amount of time together both on and off the job, we might be surrounded by this negativity 24/7.

Dealing with Complainers

So, whether for a couple of weeks or several years, how does one deal with the Tools of the fire world? What follows are a couple of suggestions on how best to handle the Tools. (Sorry, I couldn't resist.)

First and foremost, consider doing what is called a self-assessment. In other words, look in the mirror and decide if you are, indeed, a Tool. After all, before trying to change others we need to determine if our behavior needs changing first. So, ask yourself some simple questions:

- How much do I bitch out on the fireline and back at the home unit?

- Do I generally keep a pretty good attitude, or do I complain about how things are handled?

- If you were to ask someone else about my attitude, what would they say?

Be honest with yourself when answering these questions. If you're not sure about the answers, do some self-monitoring for awhile at work and on the fireline. Pay attention to how much complaining you do.

The important thing to remember is that we all bitch a little bit at one time or another. It's part of being on a crew. Complaining about the Overhead team, or the sack lunches, or your current assignment often goes hand in hand with the job. In a way, complaining about common hassles may help bring the crew closer and more cohesive as a unit. It shows that you are in it together, and all dealing with the same challenges. However, it is important to remember another old saying: Everything in moderation. A little bit of bitching is okay. Too much and you run the risk of alienating yourself from the rest of the crew and/or bringing a bunch of other people into your gloomy world of pessimism.

Fighting fire is too much fun to have to deal with constant griping. Think about it. We're outside, being active, and we are challenging ourselves in a high-risk environment. We are probably working with a great group of people, seeing some beautiful scenery, and getting paid to do it. Think about how many people would love to have the job that we have! Really, what is there to bitch about?

Jon Gruden is the head football coach of the NFL's Tampa Bay Buccaneers, who won the Super Bowl in 2003. Later that same year, he wrote a book titled, Do You Love Football?* In the book, Gruden talks about how he often gets in the faces of his players and asks them if they love the sport. Despite being paid hundreds of thousands, if not millions of dollars, Coach Gruden wants to know if they are playing because of a love for the game.

You can ask yourself the same question. Do you love fighting fire? Chances are, you could probably be doing something else for an occupation. One thing that I have found about firefighters is that they usually have some other set of skills that they could easily utilize in another career. However, they choose instead to put up with the schedule and the demands of being a wildland firefighter because they all usually share something in common: They love it! That is not to say that we aren't fighting fire to make a little money. Overtime and hazard pay are highly motivating factors in this business. However, if we are only motivated by money, we risk not getting as much enjoyment out of our work. And, when we don't enjoy our work, we are prone to soak in our PPA's.

While I was working for the New York Giants at the Senior Bowl in Mobile, Alabama, in February of 2004, I saw Coach Gruden and got him to sign a copy of the book. I told him that I had promised my son Skye that I would ask the coach to autograph it, which he gladly did. I failed to tell Coach Gruden that my son was only six months old at the time.

From time to time we firefighters may need to remind ourselves what it is that we love about our job. So, we might sit down every now and then and make a list of the things that we really love about fighting fire. Is it the people? Is it the feeling that comes from completing a difficult assignment? Is it the smell of woodsmoke on a crisp mountain morning in fire camp? Each of us finds motivation and enjoyment from different things in this job. By writing them down, we can identify what we love and clarify why they are important to us. If need be, keep this list in your wallet or tape it to the inside of your hardhat. Refer to it throughout the season, or when you find your attitude going south.

If, after conducting your self-assessment you discover that you really do not love fighting fire, then it might be time to find another occupation. And that is okay. Fighting fire is not for everybody. If you really do not enjoy it, move on and find something you do enjoy. By continuing to stay in the firefighting business when, in truth, you really do not want to be there, you run the risk of becoming a negative influence on the rest of the team. Don't be a Tool! Go find a job that you love, and give it your best.

Using "I Messages"

In review, our first step in dealing with Tool is to first determine if our own words and behavior are positive. After all, it's hypocritical to complain about another person's negative attitude if ours is just as bad, if not worse. Once you have done your self-assessment, your focus can shift to those with the PPA. How do we best deal with those on our crew with extremely negative attitudes? Well, one of the first things we can do is construct an **"I message"** for that person. "I messages" are comprised of three different parts: identifying the behavior in question, labeling an emotion that comes from that behavior, and communicating the effect that the behavior has upon you.

"I messages" let people know what impact their behavior is having on others. The opposite of "I messages" are "You messages," which basically convey the message "you suck" or "you piss me off." You messages put people on the defensive, and when this happens they tend to get emotional. Emotions can get in the way of clear communication.

In the case of someone on the crew with a PPA, an example of an "I message" might be, "I get very upset when you complain so much, because it brings down the rest of the people on the crew." Although it is a short and direct statement, it contains all three parts of an "I message." It tells the person what their behavior is, how you feel about their behavior, and what impact you think their behavior is having.

"I messages" are like most other skills: the more you practice them, the better you get at composing and delivering them. At first, they often do not feel like a natural form of communication. Our initial reaction to someone's negative behavior might be to send them a "You message," or to not even communicate with them at all (like I chose to do with Tool). But if this person's PPA is having a negative impact upon you and/or the crew, it is important that this be identified to him or her. It is also important that this come in the form of an "I message."

Supervisors and Crew Morale

But what happens if we have given the person an "I message" and their negative attitude continues? Well, it's probably time for another "I message," such as "I am frustrated because I have told you that your constant complaining is bringing down the rest of the crew, and yet you are continuing to gripe." It is also probably time for others who have been impacted by the complaining to also send the person an "I message." Perhaps this will help get it through to them that others on the crew have noticed and they, too, are affected by the negative behavior.

If their negative attitude still does not change after several "I messages," it might be time to utilize the change of command. Let your supervisor know your concerns. Perhaps your supervisor is unaware of how this person's negative attitude has affected you and the rest of the crew. Tell the leader about your attempts to communicate with the person, and what impact it has had or has not had. Resist the temptation to do nothing. Ignoring the problem will not make it go away. Crew morale is bound to be affected negatively if this person's PPA continues to go unchecked.

Of course, things get more complicated if the person with the shitty attitude is one of your supervisors. But, the same approach works here as well. Communicate with them using an "I message." Be specific when telling this person about what their behavior is, what impact it has, and how you feel about their behavior. Have others on the crew send an "I message" as well. If the supervisor's attitude does not get any better, or if he or she is unwilling to accept this feedback, go to their supervisor and tell this person of your concerns and what attempts you have made to address the issue. If you are concerned about how doing this will impact your evaluation (since the supervisor you have is probably responsible for your performance review) request that somebody else complete your performance evaluation.

Like most humans, we firefighters can have the tendency to be a gossipy bunch, at times. When we have a problem with another person on the crew, we will often discuss it amongst other crew members, but never make the person in question aware of the issue we have with them. In the case of somebody with a PPA, though, we need to communicate directly with them. This person's negative attitude will undoubtedly affect the rest of the crew at some point. By using "I messages" the person with the PPA is made aware of what he or she is doing, what impact it is having, and how you feel about it.

As we said earlier in the chapter, attitude accounts for a great deal. When it comes to attitude, the first thing to do is conduct a self-assessment. Monitor how much complaining you do. In a high-risk/high demand environment like wildland firefighting, each of us needs to work at maintaining a positive attitude as much as possible. Sure, there will be those times where you just need to vent by bitching a little bit. That's fine, as long as it does not happen on a daily basis. Now, with our own backyard in check, we can address the PPA of others on the crew. Start by sending "I messages." Follow up with more of them if the PPA does not improve. If sending more "I messages" still does not work, consider going up the chain of command to let your supervisor know of the issue and for help on how to address it. Remember, fighting fire is way too much fun to have to deal with somebody's piss poor attitude!

Goal Setting

December, 2001: One winter morning, while sitting in one of Missoula's many coffee shops, I was doing some writing. As I looked up, I happened to see Jeff Kinderman and his wife standing in line, waiting to order. At the time, Jeff was the Loft Foreman at the Missoula smokejumper base, and he has since taken over as Assistant Base Manager. We caught each other's eye, and after getting their caffeine, they came over to my table. Jeff is one of the most even-keeled, level-headed jumpers that I have had the opportunity to work with. He is respected by everyone, and has probably forgotten more about parachute canopy systems than most others have ever learned. He is, by all standards, a class act and a fine firefighter.

As firefighters are always prone to do, we began to talk about the upcoming fire season.

"What's your plan? Do you know if you're coming back to jump?" he asked. I had just recently completed my doctoral studies, and was not sure if I was going to be jumping that season or not. Eventually, I decided to jump that year, and continued to do so with the exception of 2003, when I took one season off due to the birth of my son. (That's the addictive quality of fighting fire, I guess – it's pretty tough to give up, especially when you've been doing it for so long.)

"I don't know," I replied. "It might be time for me to give it up and get a real job."

"What do you mean, 'a real job,'" Jeff asked. I could see that I had kind of pissed him off. Knowing how laid back he is, I could instantly tell from the tone of his voice that he was not too happy

with my choice of words. I immediately felt sheepish and stupid for my comment about a "real job." Here was a man who was well on his way to thirty years of government service as a firefighter. Jeff is a fantastic fireman, and a fine leader of others. He has been instrumental in the design of new parachute canopy systems and many of the design revisions that have occurred over the last several years. He has devoted most of his occupational life to firefighting, and then he has to listen to me comment on the fact that I don't think of it as a 'real job.' Needless to say, I have never made this mistake again.

So what does this exchange between Jeff and me have to do with goals, you might be asking? Well, I believe that it highlights a difference between myself and many other firefighters when it comes to goals. For me, firefighting has never really been a career, but rather just a way to financially support myself and hang out with some really good people. Jeff might very well enjoy these same aspects of the job, but he has also, I believe, embraced firefighting as a professional occupation.

Early on when I worked at the district level, it was the perfect job to have while I went to college. I could work hard for almost four months (things were under the quarter system then, not semesters), make a bunch of dough, then go back to school. A few years later, while employed as a school psychologist, I still had my summers off, which allowed me to fight fire for about three months of the year. Doing fire suppression work was like a vacation compared to testing kids and writing reports all day long.

Once I started jumping, I quit my job as a school psychologist and worked seasonally since I did not have an appointment. When my allotted 1039 hours for the year were used up, I spent the rest of the time fishing or snowboarding, hoping that my bank account would hold out until I could start working again. Eventually, I received an appointment, but by that time I had decided to go back to school and finish my doctoral work. Once again, I found myself working a shortened three-month fire season because of my academic schedule.

In other words, I have treated fire more like a hobby than an occupation, even though I have finished eighteen seasons as a firefighter. By doing so, I have passed up or failed to take courses and other training opportunities that would have helped me in the long run. While I have been pretty good at setting goals for myself in some of the other aspects of my life, I have not done the same for my work as a firefighter. I believe that if I had set some specific goals for myself each season (such as red card qualifications or S classes to pursue) that it would have made me a better firefighter. Also, setting some fire-related goals would have made me more marketable when applying for certain jobs in fire. Now, looking back, I dearly wish that I had done a better job of goal setting. Hopefully, you have. If not, it's never too late to start!

Personal Goal Setting

Decades of research encompassing hundreds of studies have demonstrated the effectiveness of goal-setting. With athletes, business professionals, students, or just about anyone else, setting goals has been shown extremely useful and beneficial for a variety of people and situations. And it does not have to be limited to occupational goals. Goals can be set in just about any aspect of a person's life: socially, academically, spiritually, or physically.

So how does one get started setting goals? To begin with, they must be consistent with an individual's hopes and dreams. There is no sense setting goals in areas that are unimportant to you. Focus on the things that really matter to you, whatever they might be, and set some goals accordingly. Next, each of us needs to discover our own goal-setting style. While some prefer to set quite a few goals, others are more comfortable with a small number (maybe even just one) of more general goals. If you are very organized and like to make lists, you will probably prefer establishing a number of detailed goals. On the other hand, if you are like me (kind of disorganized and not a real "list" kind of person), you might opt for a smaller number of general goals. Lastly, the most important thing is to write your goals down, whatever they might be. You can post them on your wall or tape them in your hardhat, or choose another way, but write them down. Then, look at the goals frequently and chart your progress toward achieving them.

One of the most common tools to use in setting goals is the **SMART** acronym, which stands for **Specific, Measurable, Attainable, Relevant,** and **Time-phased**.

- Goals are more effective if they are specific. To establish a goal for yourself of "becoming a better firefighter" is not nearly as useful as setting a goal of "getting redcarded as a Crew Boss." Since the goal of a red card qualification of Crew Boss is very specific, a person can figure out what courses and experiences are needed to obtain this rating, then work toward it.

- Measurable goals also tend to be more valuable. "Getting in better shape" is not nearly as potent as "Finish the pack test in under forty minutes." Knowing that you want to complete the three-mile test within this time limit, you can do timed practice trials and monitor your progress. You can measure, with the help of a stopwatch or wristwatch, just how you are doing at accomplishing this goal.

- Attainable refers to how possible it will be to achieve the objective that has been established. A person can set a goal of "becoming the Chief of the Forest Service," but if he is currently a GS-7 Engine Captain ten years from retirement, that goal is not very attainable. However, if this same person sets a goal of becoming an FMO (Fire Management Officer) within that timeframe, that is much more within reach.

- Relevant has to do with prioritizing the things in a person's life that are most important at a given period of time. There is no sense in focusing on professional goals if your personal life is spiraling out of control due to strains on an important relationship or financial concerns. If that should be the case, priority needs to be given to your personal issues and goals need to be set here, if appropriate.

- Time-phased is the final component of the SMART plan. What it does is simply add a time element to the goals that have been set. Goals are often described in terms of short-term (usually days), intermediate (weeks and months), and long term (years) time frames. Short-term goals would be things that you want to accomplish in the very near future, such as "by Friday I want to have completed x, y, and z." Intermediate goals are those objectives a little farther out time-wise: "At the end of this fire season, I would like to have finished revising the SOP manual." Long-term goals are very future focused, and might be objectives to accomplish years down the road, such as "by 2008 I want to be Division Group Supervisor qualified" or "by 2010 I will own my own home."

Think about the daily "to-do" lists that many of us utilize. The items on these lists are nothing more than goals that have been written down, incorporating the SMART criteria. See, you've probably been doing it all along and did not even recognize it! The list is specific, in that it clearly states what you want to accomplish. It's measurable, because you can check things off as you take care of them, and the items listed are attainable because you would not put them on the list if they were not. They are also relevant, since you have determined that each of them is important enough to include, and from a time perspective, the list is of things you want to accomplish on that day.

Now, all you have to do is comprise this same kind of "to do" list of your general goals. By writing them down and using the SMART criteria as a template, you will have a much better chance of accomplishing them. If you are already setting goals for yourself, congratulations and good luck at achieving them. If you are not, then no better time than the present to get started!

Team Goals

Just as individuals set goals for themselves, we can go through a similar process of establishing goals for the fire teams that we work on, be it a Type II crew, an IMT (Incident Management Team), a hotshot crew, engine module, rappel crew, rookie smokejumper class, or whatever crew you happen to be on for that season. A facilitator will need to be identified to help lead the process, and this can be either a formal leader of the crew (such as a superintendent, foreman, or crew boss), or a regular crewperson who feels comfortable enough with the process and is respected enough by others on the crew to elicit their support.

To begin, as a group, ask yourselves some questions. What do we want to accomplish as a team this season/fire assignment? What do we as a group value? At the end of this fire season/fire assignment, where do we want to be as a team? These questions will help direct the discussion toward the topic of team goals. Referring to your crew mission and vision statements (assuming you have them) can also be useful when trying to determine team goals. Use of the SMART criteria will help to ensure that what you come up with will be specific, measurable, attainable, relevant, and time-phased. And, just as with individual goals, write them down and keep them visible so that the team can refer to them constantly.

Team goals can be set in any or all of the three performance areas:

- physical (pack test times, group runs or physical training, minimum skills and abilities tests for every team member, and so forth)

- psychological (team stress-reduction activities, communication amongst team members, and so forth) and

- environmental (maintenance of a crew webpage or journal, inclusion of family and friends in crew activities, and so forth).

One key thing to remember for a team is to set goals that are within your control as a team. A team goal of 700 hours of overtime, for instance, might be unrealistic because it is difficult to predict how busy the fire season might be or the number of assignments the crew will receive. Next, it is important to provide the opportunity for feedback and input from all crewmembers, as this will better ensure that the group buys into the concept of team goals. Lastly, refer to your team goals throughout the course of the season or assignment to see how you are doing at accomplishing the goals.

Years and years of research have consistently shown the benefit of goals. Those who set them are much more apt to find success than those who do not. As firefighters, we can establish both individual goals, as well as goals for the teams that we work with. Goals can be set in the areas of physical performance, psychological factors, or environmental arenas. By setting goals, firefighters can better direct their focus, and increase the probability that they accomplish those things which they have identified as being important. Through the use of the SMART criteria, goal setters can be more confident that their goals will be reached. The end result of effective goal setting: better overall performance, not only for individuals, but also for the teams they form.

8
Teams

June 5, 1995: Thirty-four of us have reported to rookie smokejumper training in Missoula, Montana. Looking around, I recognize only one face, that of Rick Lang. We each have left behind our firefighter jobs at Bureau of Land Management based out of Miles City, Montana, both of us hoping to make it through the arduous five weeks of rookie training. Surrounding Rick and me are others who are forsaking guaranteed jobs on their hotshot, heli-rappel, or district crews for the chance to earn the wings of a smokejumper. Handshakes all around, but I remembered few of the names of people to whom I was introduced. There are simply too many faces, and I am too nervous to think straight.

After the PT test (pull-ups, push-ups, sit-ups, 1.5 mile run) our number is reduced, thinned of a few individuals who are unable to meet the minimums. The trainers keep telling us that we had better be prepared to do a whole lot more, physically, than what the PT test demands. After gathering up gear, we load into vans and head into the mountains for the first week of camp. I keep to myself, alone in my thoughts and self-doubts about whether I have enough strength to make it through the training.

In the rain, we set up the wall tents that will be our homes for the next four nights. I begin to learn more about some of the individuals around me: Ed from the Avery, Idaho district, Ted from the Helena Hotshots, Steve off the Flagstaff 'shots, Brian from Sled Springs Heli-rappel. That night, rain turns to snow, and when we waken for the first of several long runs, we must slog through a foot of the slushy white stuff. Next comes our first of many grueling line digs. Hands and minds become numb because of the cold. When one of us begins to slow, others are there to provide encouragement.

For three-and-a-half more days and nights it continues, my body getting beaten down by the weather, a lack of sleep, and the physical demands. A hundred times I tell myself that it's time to quit, but somehow I resist the urge to do so. After the 85 pound pack test (a grueling two and a half mile climb/descent/climb over nasty terrain), I vomit a partially digested MRE into the snow and retire to my cot. I waken at times to hear the others laughing around a campfire, eating the first hot meal provided all week. I am too sick to get out of my thin, yellow sleeping bag. I pull on another sweatshirt (the last dry one that I have) and try to quell the uncontrollable shivers that overtake my body. Sleep, reluctantly, comes. A few of my rookie brothers stick their heads into the tent to see how I'm doing, but I am dead to the world.

Friday, we break camp and head back to Missoula for the 110 pound pack test. The early finishers come back to cheer us slower ones on. Every one makes it within the time limit. With the test complete, our first week comes to a close. That weekend we all spend time together, limping around on blistered feet, drinking beer, and talking about the week of jump-training that will begin on Monday. We're out of the woods literally, but far from out of the woods in the figurative sense. We still have a long way to go to earn our jumper pins. Everyone says it just gets harder from here. Slowly though, we are molding into a powerful team.

Wildland firefighting, be it the initial attack of a small remote fire or the suppression efforts undertaken on a huge Type I complex, relies on teams of people to accomplish the mission objectives. Therefore, how we function as a team will largely determine how effective and safe our efforts will be. Often times, the teams we work with will be established at the beginning of fire season, and will remain relatively constant throughout the course of the year, such as a 20-person Type I Hotshot Crew, an engine module, a Helitack Crew, or an Incident Management Team. These "long-term" teams allow us to work with the same group of people on a near-daily basis. With such frequent and lasting contact, a team is gradually formed. As members begin to know more and more about each other, trust builds, and with it cohesion.

Other teams are brought together on short notice, and are expected to perform well in a short amount of time. This includes some Type II crews (such as American Indian firefighters), or district crews that are combined with other district crews in order to obtain the 20 people needed. Other "short notice" teams include members taken from several different teams to accomplish a certain specified mission. An example of this would have been the group formed at South Canyon when part of the Prineville Hotshots combined with a mixed group of smokejumpers on the West Flank fireline. "Short notice" teams present unique challenges to the leaders in charge of them, as will be discussed in greater depth later on. Long-term teams are not without their own set of challenges, and these too will be addressed.

What is a Team?

Simply defined, a team is a group of people working together in a coordinated effort on a common mission. As firefighters, the goal of our teams is often the suppression of fires. However, as our job duties expand, our teams are becoming responsible for a variety of other tasks, be it prescription burning, various project work, or missions that fall under the general heading of "all risk," such as searching for shuttle parts in Texas, killing flu-stricken chickens in California, or assisting in hurricane clean up efforts such as Katrina and Rita. Clearly, the more effective we are working in a team environment, the better we can respond to the multitude of different tasks to which we are assigned.

Cohesion and Teams

In order for any group to work together effectively, efficiently, and safely, many different factors must be present. One of the most important factors is the level of cohesion within the group. In the sports world, one constantly hears about the importance of teams "bonding" or "gelling" or "having good chemistry." The Montreal Canadians in hockey, the Pittsburgh Steelers in football, the New York Yankees in baseball, and the Chicago Bulls in basketball all serve as historical examples of just what cohesive teams can do. Few would doubt the positive effects that can be generated when a group of individuals works together collectively as a team.

The topic of cohesion has been, and continues to be, one of the most studied areas in the field of human behavior. While many definitions of the term cohesion have been offered by various researchers, the one most widely used and accepted in the field of sport psychology was developed over fifty years ago. Festinger, Schacter, & Black (1950) defined cohesion as "the total field of forces which act on members to remain in a particular group." In our case, the group would be wildland firefighters. So the question becomes just what are these "forces" that bind us together?

Many scientific studies of group behavior have revealed that group cohesion is not one thing. Many different aspects of "group cohesion" have been identified and discussed in the literature of psychology. One of the most important distinctions that has been studied is between social cohesion and task cohesion (MacCoun, 1996). According to MacCoun, social cohesion refers to the nature and quality of the emotional bonds of friendship, liking, caring, and closeness among group members. A group displays high social cohesion to the extent that its members enjoy each other's company, choose to spend their free time together, and feel emotionally close to one another. Task cohesion, on the other hand, refers to the shared commitment among members to achieving a goal that requires collective efforts of the entire group. A group with high task cohesion is composed of members who share a common goal and who are motivated to coordinate their efforts as a team to achieve that goal.

Having pinpointed these two types of cohesion (social and task), certain questions arise. Is one type more important to firefighting than the other; and, if so, which one? And, since cohesion on the fireline is important, how do we go about developing and maintaining it within the firefighter community? According to MacCoun (1996), task cohesion may be more important than social cohesion in enhancing group performance. After reviewing studies which investigated cohesion and performance in both military and civilian groups, MacCoun concluded that it is task cohesion, not social cohesion or group pride, that drives group performance. So, as firefighters, we should be particularly concerned with task cohesion, since it propels group performance, and outstanding group performance is what we need in our jobs. One should not conclude that we should ignore social cohesion. Rather, since our safety depends upon winning a battle against wildfire (or whatever "all-risk" mission we happen to be doing), like all soldiers on the front line we must first and foremost develop high task cohesion.

How then, do we go about developing and maintaining cohesion on the firefighting line? What we must remember is that we may have done a great job of becoming a cohesive workforce, but room for improvement always exists. The following represents a list of some general suggestions about how to increase levels of task cohesion:

- Conduct a briefing/debriefing before and after every fire, even on the smallest fires. As firefighters, we learn how to do our job better during briefings. Unfortunately, briefings/debriefings are not done on every fire. In a good briefing, we outline the mission, provide intent, assign tasks, foster communications, develop LCES (Lookouts, Communication, Escape Routes, and Safety Zones), and answer questions about a host of other things. Groups with high task cohesion are motivated to coordinate their efforts as a team to achieve a goal. With briefings, we build task cohesion. Also, briefings/debriefings can be done for project work and other missions, too. Once the mission is completed, a thorough debriefing or After-Actions Review (AAR) allows us to talk about the intended objectives, what happened, why it happened, what we could have done differently, and what things worked well (AAR format is included in Incident Response Pocket Guides).

- We must also get into the habit of getting to know the people we work with shoulder to shoulder. If you find yourself working with people you don't know, take it upon yourself to learn their names, where they are from, what is their overall fire experience, and so on. Group leaders obviously will do this early on, but whether we hold a leadership position or not should not matter – getting acquainted with your group should become a rule of thumb for all of us. Knowing your co-workers will help develop social cohesion. It is difficult to establish social cohesion if you do not even know the names, the home towns, and the experience levels of the people you are working with.

- If you find yourself working on a fire with people you do not know or have never worked with before, realize the level of cohesion will probably be lower than it would be if you were all from the same crew or district, at least initially. If possible, try to avoid segregating yourself based upon crew affiliation. We tend to hang out with those people we know (there's nothing wrong with that), but always hanging out with your buddies can and does have a negative impact upon overall social cohesion when there are new faces on the fireline. Try to work alongside people you don't know if the opportunity presents itself.

- Strive for mutual respect among group members. In order to have a cohesive group, we need to show respect to our teammates. We don't need to love our teammates away from the fireline, but we do need to respect them while we are on it. Remember, the mission takes priority, not petty squabbles among group members.

- Maintain effective two-way communication that is both direct and clear. When people communicate effectively, a host of other problems can be avoided. If things should happen to "go south" on the fireline, good communication between firefighters will be even more paramount in importance. Related to this topic of effective communication, you must always know the goals and objectives of your mission. If you find yourself unsure as to what the goals and objectives are, take the time to find out.

- Realize that high social cohesion is not always a good thing. Many years ago, Irving Janis coined the term "groupthink" to identify the failure of a highly cohesive group to engage in effective decision-making processes. In this scenario, group members become so concerned with getting along that they fail to consider information which does not support the group's decisions. Members feel such a pressure to conform to group standards that they fail to bring up important questions or concerns. When this happens, disaster can result. As firefighters, we need to learn how to communicate with our leaders when we are uncomfortable with a specific mission or task. The South Canyon, Thirtymile, and Cramer fires serve as recent examples of what can happen when we do not communicate as effectively as we should within our groups.

Understanding How Teams Develop

Every group develops along a unique path, but teams also go through a series of predictable stages. By being aware of what these predictable stages are and how the team development process works, you will find yourself in better position to understand and deal with the dynamics that occur when a group of people is forming into a team. If you are a leader (whether a crew boss, squad boss, engine boss/captain, Jumper In Charge, or Incident Commander, for example) being familiar with the team development process should enable you to better lead and manage those people who work for you. But even if you are just a crew member, understanding how teams form will help you to be aware of group dynamics and the role you might play in your team's development.

Teams develop differently according to many factors, such as the number of new group members being incorporated at a time, and positions held by the new individuals. For example, a Hotshot crew or Heli-rappel crew may bring on only 2-4 new members each year. These new individuals might be crew members, or they might be members of the crew leadership structure, or they may be a combination of both. On the other hand, some crews (such as 20-person Type II's) might be composed of all or predominantly new members, and it becomes the Crew Boss's job to mold them into an effective work force. These kinds of "short notice" crews have been described earlier, and pose a particular challenge to the leader. Regardless of how many new members are being incorporated into the group, the development stages that the group will go through typically remain the same.

Five Stages of Group Development

Tuckman (1965, 1977) studied group development for many years, and he concluded that groups go through five distinct stages of group development, which he calls **Forming, Storming, Norming, Performing,** and **Adjourning.**

Stage 1: During the Forming stage, the relationships between individual members are characterized by dependence. Members strive for acceptance by the others in the group, and often will look to the group leader for direction and guidance on what they are supposed to do. Since new individuals often do not know other crewmembers or even their leaders, they engage in a process of gathering information to find out who they can trust, who they are similar/dissimilar to, and where the sources of group power reside. At this stage, new individuals attempt to become oriented both with one another and with the tasks they are responsible for accomplishing.

As noted in the Fireline Leadership (L-380) coursebook published by Mission-Centered Solutions:

The formation phase is characterized by a lack of situation awareness (your perceptions do not match reality very well). New crewmembers do not know other crewmembers or their leader. They do not know what kind of behavior to expect . . . At the start, crewmember's roles and responsibilities may be undefined. Standard policies and procedures may be unclear. Communication norms and acceptable methods for dealing with conflict may not be spelled out (p. 135).

During the Formation stage, the general rules of behavior often become keeping things simple and striving to avoid any controversy or behavior that makes you stand out negatively within the group. To transition from this stage to the next stage, new members must move from this comfort zone of "not wanting to rock the boat" and risk the possibility of conflict.

As a leader at this stage, several steps can be taken to help new crewmembers function within the group:

- Communicate clearly and effectively. Lack of information causes stress, so it becomes the leader's job to "fill in the blanks" that new members might be experiencing. The importance of early communication cannot be over-emphasized. Communication can be accomplished in many ways: by an initial letter or phone call introducing yourself and detailing your expectations of the new employee, via a crew webpage which provides useful and pertinent information, through group briefings, and with one-on-one meetings with those who are new to the organization.

- Consider assigning a sponsor or mentor to new crewmembers. This sponsor should be someone that you trust as being a good role model. This can be done in advance of the first day that the new person shows up for work. Any initial questions or concerns can then be dealt with prior to starting out, and this new person has a headstart on a new friendship once he or she shows up for work.

- Take time to reflect on what it was like for you when you started your first fire job, or moved to a new job in a new location. By remembering and visualizing this time of uncertainty in your own life, you will be better able to empathize with the new people on your crew, and better understand what questions they are facing and the answers they are looking for.

- Be careful of "initiation practices" for new hires. While some of these activities can be good-natured and even a learning tool, they can be carried too far and become detrimental to your efforts of building a team.

As a new crewmember at this stage, several steps can be taken to make this adjustment to your new job easier:

- Show up a few days or even a week earlier than you are scheduled to report. This will give you a chance to get acclimated (both physically and mentally) to your new surroundings. Showing up early provides an early "intelligence report" on people you might be working with, places of interest in the area, recreational opportunities, and so forth. Be prepared to look for a place to stay if you are relying on government- provided housing, as you may not be able to move in before you officially start work. If you roll in the same morning you are scheduled to start, you may not have the chance to mentally adjust to all the changes you are about to encounter.

- Maintain a good attitude once you do start work (refer to Attitude chapter). Firefighters, like most people, have a tendency to form quick opinions on new people they meet. Realize that some of your experienced co-workers will size you up very quickly. Volunteer for the tough jobs. Understand that, as the new person, you will at times be assigned tasks that no one else wants to do. Tackle your duties with enthusiasm.

- Be prepared physically for the demands of your new job. Nothing singles you out quicker (in a negative way) than being unable to perform the physical requirements of the job.

- Have a good pair of boots! Texas Steers from Kmart ain't gonna cut the mustard.

As a new leader coming into an already established crew or team, the following are some suggestions on how to make your transition easier:

- Meet first with the other leaders. Doing so gives you the chance to get to know one another, to begin to learn the crew values, and time to figure out the system that is currently in place.

- Then meet with those you are in charge of supervising. Share information about yourself: your background, your experience, your leadership strengths and weaknesses, your vision. Realize that you are probably being judged very quickly, so strive for thoroughness and preparedness. Establish guidelines on how they can come to you with issues, problems, concerns, and so forth.

- Identify who the informal leaders and powerbrokers are. All crews have informal leaders (those who can influence others) who are not in a formally identified leadership position. Getting these people on your side right away will go a long way toward getting the rest of the crew to buy into your plans and vision.

- If you are the new leader coming in and have won this job over somebody else who is still on the crew, understand that, at least initially, this will present a somewhat tense dynamic. This person might feel some animosity toward you for having gotten the job. You may want to talk with him or her to clarify that you did not make the selection, and that you were simply applying for the job. One thing that you both have in common is that you both want the crew to be the best that it can be. Since your goal for the crew is the same, you now need to work together to make that happen.

- Try not to change too many things early on. This can send the message that you think the crew was "broken" before you showed up and that it needs immediate fixing. People are often resistant to change, especially when someone who is new comes in and starts messing with "the way stuff has always been done." Naturally as a leader, you will have some ideas on how you want to improve the present system. Take some time and identify a list of things you want to address or change. Do your own version of triage, and figure out which of these is most important, then the things that are next in importance, and so on. You might want to work on some of the smaller things before you tackle the huge issues (unless of course the present Standard Operating Procedures are unsafe or unhealthy to crewmembers).

Stage 2: This stage, which Tuckman (1965, 1977) calls the Storming phase, is marked by competition and conflict among group members. As individuals become more familiar with one another, the formalities and niceties that existed during the Forming phase give way. People's "true colors" begin to show. Since crewmembers are now more apt to take risks and challenge one another, conflicts between individuals or groups arise. These conflicts are often the crew's way to determine who has power and authority within the group (both formal and informal). These conflicts may be out in the open, visible for all to see, or they may just simmer "below the radar screen." Either way, the conflict is there. Members may display wide swings in behavior due to these emerging issues of competition and conflict. Since this stage generates discomfort among the members, some may keep it to themselves and remain quiet, while others attempt to assert their dominance.

The important thing to remember is that the conflict generated during this stage is completely normal and natural.

As crewmembers attempt to understand just what their role is on the crew and what their responsibilities are, there will be some head-butting. In order to transition from the Storming phase to the next phase, members must move from a "testing and proving" mentality to a "problem-solving" mentality, where the needs of the crew outweigh individual differences. The key to moving on to this next stage is communication: how well members talk with and listen to one another. If individuals have not been able to effectively deal with the conflict that arises during the Storming phase, the crew will stagnate and true cohesion will not develop. The group will either implode into itself, or will continue to function in a sub-optimal way.

As a leader, some of the following steps should help:

- Clearly define the chain of command (COC). Make sure individuals know who the formal leaders are, and emphasize working through the COC when conflict arises.

- Hold frequent crew briefings. Talk about how it is normal for conflict to arise during this part of crew development, and give people a chance air out their feelings.

- Emphasize teamwork all the time. Establish team goals (see Goal-setting chapter) and encourage people to make decisions that put the team first.

- Focus on the "end state", which is your vision for the team. Realize that if the team continues to stay in the Storming phase, it will be difficult to accomplish what it hopes to do. The "end state" should be a healthy and productive team.

- Lastly, not all conflict is bad. Unhealthy conflict focuses on *who* is right, while healthy conflict centers around *what* is right. Ineffective teams allow personal squabbles and differences of opinion among members to become roadblocks for team performance. Highly effective teams are able to manage healthy conflict as members work through issues and challenges. The difference is that healthy conflict is productive, and members have enough trust with one another to not take it personally, and are able to focus on what is right for the team.

Stage 3: During the Norming stage, trust finally begins to develop within the crew. Group members actively accept all the contributions and input from one another. They begin to act out the old cliché: There is no "I" in team (although there is an "m" and an "e," but that's a topic for another day). Cliques that were formed begin to dissolve away. Since individuals are now able to trust one another, this leads to the development of crew cohesion. With interpersonal conflicts resolved, members feel a sense of relief and begin to experience a sense of group belonging.

Stage 3 is marked by a flow of information between members. With feelings and ideas shared, crewmembers provide feedback to one another, and the group becomes better able to focus on their common task or mission. If a group is able to reach this stage, members begin to feel good about being part of an effective team.

Leader tips:

- If you see your crew getting to this stage, it is probably also well on its way toward the next stage, Performing. No major modifications are really necessary at this point. The key challenge at this middle stage in team development is getting members to commit (and keeping them committed) to the concept of team. Whenever possible, reinforce behaviors and attitudes that place the team ahead of individual concerns.

- If your crew is still not at this stage after a reasonable period of time, go back to the tips provided under the Forming and Storming sections. Make adjustments as necessary. Chances are the team has become stuck in the Storming phase, and unresolved conflicts are keeping them from moving into the Norming phase. Conflicts need to be addressed and resolved if the team wants to develop.

Stage 4: The Performing stage, where the team is at or near optimal performance, is not reached by all groups. If the crew has been able to get this far, it achieves a state of grace. In stage 4, members can work effectively alone, in small subgroups, or as a total team unit. At the personal level, they display interdependence with one another. At the task level, they actively problem-solve to accomplish missions. The crew is both people-oriented and task-focused. In other words, the group has both high social cohesion and task cohesion. Unity within the group is displayed: morale is high and loyalty to the crew is profound. Synergy is attained, where the product of the group is greater than the sum of its individual members. The fireline gets cut, trails get built, things get done, and crew members have a good time doing it. The team functions like a well-oiled machine.

Leader tips:

- The Performing stage should be your goal for every team that you are associated with. If you have not reached this stage, analyze the crew and figure out what stage they are presently at, then take the steps to get them to this level.

- If you are consistently able to get your team to this level, be proud of yourself. Keep notes of your ideas and techniques on how you are able to do this. If times get tough, sell these notes to the highest bidder.

Stage 5: The Adjourning phase occurs when the mission for the crew has ended. With a "short notice" team, this generally happens when it has completed its assignment or has reached the 14-day limit. For longer-term teams, Adjourning takes place at the end of the fire season. Terminating a group can create new challenges for its members, especially if the group has been effective and the individuals have formed strong bonds with one another. Some times the group will terminate as one entity (typically with Type II crews), but at other times individuals will cease their employment at varying times due to personal schedules. As a leader, some of these tips might help during this Adjourning process:

• Do a close-out team AAR for the specific mission/season. Have someone take notes. Talk about the objectives that you had. Discuss what happened during the mission/fire season, why it might have happened, what you can improve upon for next time/year, and emphasize the good things that the crew accomplished.

• Make every effort to meet individually with the members, even if only for a short time. Obviously, you will have to meet with those you supervise in order to do his or her evaluation. But this one-on-one meeting also provides a forum for discussion on a more personal level. Having this meeting communicates to the employee that you care about him or her as an individual and are interested in hearing what they have to say. Ask what things you, as a leader, can do to improve.

• Recognize those who did great work. This recognition can be formal or informal. It can come in the shape of cash awards, plaques, certificates, team photos or apparel, or simply a pat on the back.

For crewmembers, here are suggestions for the Adjourning phase:

- Be prepared with questions, comments, and feedback for your leaders during your individual meeting with him or her. If a one-on-one meeting with your supervisor has not been scheduled, take it upon yourself to set one up. Provide the leader with your individual perspective on how the mission or year went and how the team functioned.

- Keep copies of any important paperwork that you might receive.

- If you still have not bought a good pair of boots, take some of that "OT" money and head for the boot shop!

Now, in summarizing the five stages of team development, we can say that after a group forms friction will inevitably result as individual members try to figure out their specific role and the hierarchy of power. As members become comfortable with their place on the team, they begin to think more about the group than of themselves individually. That's when the team is in position to effectively perform the mission or task that it has been given. Lastly, the group dissolves, and members move on, possibly to form new teams. By understanding these predictable stages of team development, we can improve our functioning within that environment, whether we are a leader or a crewmember.

Team Building Advice from Coaches

Sport psychologists have long been fascinated with team-building. As team sports have become more visible and significant at the professional, college, and Olympic levels, many psychologists have worked toward finding ways to improve the functioning of teams. The results of these efforts apply not only to elite athletes but also to firefighters.

Yukelson (1997) has recommended seven suggestions for helping coaches impact team functioning. In many ways, those with leadership positions in fire — superintendent, crew boss, smokejumper or Hotshot foreman, FMO (Fire Management Officer), or AFMO (Assistant Fire Management Officer) — are much like coaches in sport. Therefore, Yukelson's suggestions for coaches should be applicable to fire leaders as well. His list includes:

1. Get to know team members as unique individuals

2. Foster pride in group membership and a shared sense of team identity

3. Develop a comprehensive team goal-setting program

4. Provide for goal evaluations, and set aside time to review how successful the team was at accomplishing its goals

5. Clarify role expectations for individual members of the team

6. Set aside time for team meetings

7. Establish a player counsel (one person who acts as spokesperson for team concerns or issues)

With this list identified, some elaboration on each of the seven strategies for athletes would help relate what they do to our work as wildland firefighters.

Get to know firefighters as unique individuals: Although firefighters might tend to look alike dressed in green Nomex and yellow shirts, in actuality each and every one of us is unique. Phil Jackson, the current coach of the Los Angeles Lakers, has written several books focusing on leadership and team issues. He frequently emphasizes that since each person is different, leaders cannot use a "one size fits all" mentality when working with team members. By knowing the individual traits and strengths of each individual, those in charge will better know how to motivate and lead those on the team. In order to do this, leaders must spend one-on-one time with each and every team member. While it will take commitment on the front end to do this, in the long run time will be saved because of this positive, early relationship that has been formed.

For individual members, they must also take the time and energy to become familiar with their teammates. It is not possible to develop cohesion without knowing who else is on your team. Some crews will have the luxury of spending all season together, so the chances to learn about one another will be frequent. Other groups, though, will be together for only a short time. This decreases the window of opportunity to get to know one another. Members must assume the responsibility of learning about their teammates. If individuals do not take the initiative to get better acquainted with one another, leaders may have to facilitate the process.

Develop pride in group membership and a sense of team identity: For a true team to develop, individuals must place their own needs secondary to the needs of the team. By doing so, behaviors and actions are guided by the principle of "what is best for the team" instead of "what is best for me." Therefore, leaders and team members alike must strive for this "Team First" attitude.

A great first step for many teams is simply having members wear matching crew apparel (shirts, hats, coats), which in effect become the "uniform" for the group. Real team spirit goes deeper than just everyone wearing the same clothing, however. To truly develop pride in group membership and a sense of team identity, leaders and followers alike will have to be willing to go much further. Team photos of the crew can be taken and prominently displayed. A written history can be done to document the crew's early beginnings and development, and members should be familiar with this history. Those on the team might be encouraged to keep journals, which highlight the day to day activities of the crew. These can be made available to future members. Former firefighters on the crew might be brought in to discuss what being a team member has meant to them. The list of possible activities is long, and it should be up to the current members to decide what things they want incorporate in the effort to build team identity. Team-building is an active process that takes planning and persistence.

Develop a comprehensive team goal-setting program: This has already been covered in the Goal Setting chapter, but it is definitely worth mentioning again. Regardless of whether it is a long-term or short-notice team, team goals can help to guide the expectations and behaviors of the members. What does the team want to accomplish? What does "right" look like? Goals can help to build the team's identity.

Provide for goal evaluations: Simply setting goals for the team is not enough. Time must also be set aside to review how successful the team has been at accomplishing these goals. At a minimum, this should be done at least two times: at the mid-point of the season or assignment, and at the end. If at the mid-point evaluation it is determined that the goal or goals are not being met, then either the objectives can be modified or the team can renew their commitment to meeting them in the time that remains. At the end of the season or assignment, the team can evaluate how they did at meeting the goals. For those that were accomplished, the team can discuss how they were able to meet them. If there were goals that were not accomplished, members can discuss what kept them from doing so.

Clarify role expectations: Each and every member of the team has a role, from the brand-new rookie all the way up to the experienced veteran in charge. By making sure that everyone knows that their contributions are important, members can begin to see how they fit into the total scheme of team functioning. However, these roles must be defined so that individuals know what is expected of them. This does not have to be the sole responsibility of crew leaders. Experienced team members and informal leaders can help in this process as well.

Set aside time for team meetings: The fire community has gotten much better at utilizing debriefings and After Action Reviews (AAR's). In essence, these are really just team meetings with an emphasis on a specific incident or event. So, if your crew or team has been doing debriefings and AAR's, the group is well on its way. Regular team gatherings can also focus specifically on how well the group is functioning. If things are going well, team meetings allow for discussion of what is working and why. However, teamwork is not always easy. Challenges will arise which threaten the integrity of the unit. By having time set aside to address emerging issues, the team as a whole can discuss what changes need to be made.

Establish a player counsel: Regardless of how gelled or bonded a team may be, there almost always will be some barrier between those folks in leadership positions and the regular members of the team. This is just part of the team dynamic. However, if these barriers become too high or impenetrable, problems can arise in team functioning. A way to address this is to identify one person who acts as spokesperson for team concerns or issues. In most instances, it should probably be a respected member of the crew who is not in a formal leadership position. This way, regular members of the team will have a formal conduit to the leaders of the crew. The team counsel can act as a bridge, helping to join the leaders with the led. Many topics can be addressed during the regular team meetings. However, in those instances in which members might not feel comfortable raising the issue within this group session, the team counsel can be utilized.

Since many of these seven strategies to better team functioning seem to be common sense, most leaders are probably already doing their own version of some or all of the things on this list. If so, this should validate that your approaches as a fire leader are shared by successful coaches and athletes. If some of the items on the list are unfamiliar to you, perhaps incorporating them into your present "leader's toolbox" will improve your effectiveness.

As firefighters, we rarely work alone. In one way or another, in almost all cases, each of us is simply one member of a larger team. Therefore, effective teamwork and team functioning are vital to the success of our missions, whatever they might be. Being familiar with social and task cohesion will help to clarify how each relates to the growth of the team. An understanding of the five stages of team development should lead to a better grasp on the processes that are involved whenever a group of people come together. And, by knowing some strategies to improve team functioning, both the leaders and the led will add to their respective arsenals.

9
Stress

June 21, 1995: The day of our first jump has finally arrived. After nearly three weeks of training, we are ready to put our rookie parachuting skills to the test. This first jump was supposed to have taken place two days ago, but rain postponed those plans. The forecast today calls for sun, so everything should be a go. I have been a nervous wreck for days, dreading and yet at the same time eagerly looking forward to getting this first jump behind me. I have been unable to sleep much at night, my thoughts and fears of injury and death overcoming my body's desire for rest.

Twelve of us board the Sherpa, and after pre-flight checks, we take off into the cloudless sky. We will jump one at a time into an open field at the Blue Mountain Recreation Area just outside Missoula. When my turn comes, I barely hear the spotter's briefing over the wind blast from the open door. My mind is a jumbled mixture of adrenaline and paralyzing dread. When the slap on my calf comes, signaling my time to bail out, I flop out the door. Immediately my legs get blown up over my head due to hitting the aircraft's slipstream. The canopy opens with a crack, and my body snaps back into a feet-down, head-up position. Instructions from trainers on the ground come over the King radio that I have in my pack, but I barely hear what is being said to me. I am definitely on sensory overload at this point.

One hundred feet above the jump spot, I for some reason decide to pull my toggles all the way down to my waist, even though this is not how we were trained to land. This brings the canopy to a stop, and I begin to drop quickly as the air spills out of the parachute. Before I know it, the ground is rushing at me in a blur. I hit with tremendous force, my left ankle turning under the weight of my body as it attempts a pathetic parachute landing fall. I get up and quickly yell, "Palmer, okay!" even though I know it's a lie. My ankle throbs, and I fear that this first jump may have also been the last. We congratulate one another and snap a few photos, then hike back to the rigs. I try to hide my limp, but it's no use. Several of the trainers have seen how I landed. They know I'm hurting. Hopefully, I'll be able to get it iced down tonight, and taped up for the next six training jumps. I need to keep my mind and emotions in check if I have any chance of making it through.

What is Stress?

In a nutshell, stress is an involuntary psychophysiological reaction to any demand placed on the body, which requires an adjustment. (Repeat your last, over?) Basically, stress impacts us all, in mind and body. Situations that are stressful trigger an inborn response that has survived millions of years — the "fight or flight" reaction. This byproduct of evolution has many effects: it increases blood pressure, heart rate, respiration, blood flow to muscles, and metabolism, all in preparation for conflict or escape.

This "stress response" is natural, and can be beneficial in that it primes our bodies for action. Stress-related problems arise when this response is prolonged or triggered too often without sufficient adjustments to counter its effects. Lazarus (1984) has argued that stress begins with a person's appraisal of any given situation. According to Lazarus, people ask themselves how dangerous or difficult a situation is and what resources they have to help cope with it. Anxious, stressed people often decide that an event is dangerous, difficult, or painful, and that they do not have the resources to cope. The key, then, is to avoid being one of these people. More on this later.

According to some researchers (Davis, Eshelman, & McKay, 1995), stress is experienced from four basic sources:

1. Your **environment** bombards you with demands to adjust: bad weather, noise, traffic, and pollution.

2. You must cope with **social stressors** such as deadlines, financial problems, job interviews, presentations, disagreements, demands for your time and attention, and loss of loved ones.

3. A third source of stress is **physiological**. Rapid growth during adolescence, menopause in women, illness, aging, accidents, lack of exercise, poor nutrition, and sleep disturbances all tax the body. Your physiological reaction to environmental and social threats and changes can also result in stressful symptoms such as muscle tension, headaches, stomach upset, and anxiety.

4. The fourth source of stress is **your thoughts**. Your brain interprets and translates complex changes in your environment and body and determines when to turn on the "emergency response." (p. 1).

Firefighters and Stress

As we all know, our jobs can be very stressful at times, with some occasions or events being particularly challenging. With these four sources identified, let's take a look at our profession with regard to how we might be impacted by stress:

The Environment: As members of a high-risk profession, each of us must contend with the reality that we can be seriously hurt, or even killed, while on the job. Our working environment is an extremely hazardous work zone: the fire itself, the terrain, weather, smoke, snags, aircraft, motor vehicles, chainsaws, heavy equipment, wild animals, insects, poisonous plants — the list goes on. All these add up to one stress-filled place.

Social Stressors: Not only is our work environment filled with hazards, but so too is our social environment. This includes stressors both on and off the job. At work, we must deal with deadlines, budgets, personnel matters, downsizing, and in-baskets that typically far outweigh out-baskets. We often work long hours and are gone for extended periods of time. Work then builds up at the office and relationships on the home front can become tense. Things that need doing (time spent with loved ones and friends, paying bills, cleaning, maintenance, and so forth) tend to get neglected when fire season is hot.

Physiological: As mentioned many times in this guide, wildland fire suppression is extremely taxing upon the body. Poor nutrition, little sleep, and long shifts often become the norm during fire season. Illness and injury can occur as the body's defenses get broken down. Physiological stress may be the most common kind we firefighters experience, especially during active fire seasons.

Our Thoughts: When it comes to stress, the mind is basically the Incident Commander on scene. While simultaneously monitoring both the environment and the body, the brain interprets and translates complex changes in both places. Just like an IC, the mind tries to gather information before making a decision. If the brain determines that a situation warrants turning on this "emergency response," it does so. In other words, how we think and what we think about can produce stress.

As firefighters, we can experience stress from all four sources simultaneously! So, what can we do about it? Quite a bit, actually. The first question is to ask: Is it possible to change this situation that is causing me stress? Your answer to this question will determine what initial steps you take. If you have the ability to control or change a stressful situation, then you can take action to address *what* it is that is stressing you out. However, if the stressful situation is beyond your control, then you must focus on changing the *way* you think about or react to this event that brings stress.

A couple of examples may help to clarify two kinds of stress: sources **within our control** and sources **outside our control**. An example of the first kind is "Tom." On the job, Tom works very hard and is asked to do more and more. Since Tom is so diligent and efficient, people end up bringing things to him because they realize that he will get the job done. Often times, these tasks aren't even within Tom's area of expertise or even in his job description. However, Tom has a hard time saying "no" to anyone, so the pile just keeps on growing. After awhile Tom may have trouble getting his required work done because he takes on so many other projects. Quietly, he starts singing an old Kenny Loggins tune to himself, the one from Top Gun…"Highway to the danger zone, highway to the danger zone…" Realizing that he is under a great deal of stress, Tom asks himself, "Can I control the situation?" The answer: yes!

Much of this stress on Tom is well **within his control.** The key here lies in being assertive with these people who are dumping work on him. Since Tom likes to please, though, it's difficult to say no. However, doing these other tasks is keeping him from getting his work done, and it's causing him stress. Time to put the ol' White boot down. Time to use an "I" message.

Tom needs to begin by being honest with people. He should communicate clearly what's going on for him because of these additional tasks. He can tell them he appreciates the fact that they are coming to him. However, this extra work is keeping Tom from getting his required stuff done. It is time for him to take his cue from Nancy Reagan and "Just say no" to them. It's not as hard as you think. It might sound something like this: "I appreciate you thinking of me when something needs to be done, but I end up feeling frustrated by this extra work because it keeps me from getting my assigned tasks done, and I end up feeling stressed out over the whole deal." Not so bad, huh?

The other type of stress that we can experience is the kind over which we have **no control**. This could be felt by Tom due to the death or injury of someone he was close to. It might be weather or traffic jams or a slow fire season. Whatever it is, the key is that we have little or no control over the basic situation producing the turmoil. However, what we can control is how we react to the event. To the body, stress is stress. It does not care if you can control the situation or not. Therefore, by controlling how we **react** to the situation, we can reduce stress. If you can keep your mind relaxed, the body will follow.

As mentioned earlier, some people are more prone than others to interpret situations as stressful. We all know them. These are the people who seem to be stressed out all the time, or who get stressed very easily.

The opposite of the chronically stressed-out people are those individuals who do not seem to be stressed out by anything. We usually know folks like this, too. They are the "ice water in the veins" type. According to Kobasa, et al (1985), these "stress hardy" individuals are less prone to illness and missing work due to sickness. Instead of viewing stressors as threats, they are seen as challenges and opportunities for personal growth. Rather than seeing themselves as having little or no control over what happens to them, these people believe that they can influence the events that happen around them.

The Biology of Stress

Hans Seyle (1907-1982) was a Hungarian born biologist who did a great deal of research in the area of stress. He is credited with writing 33 books and over 1600 scientific articles during his career, the most famous probably being his book *The Stress of Life*. Seyle developed what he called the General Adaptation Syndrome, otherwise known as the G.A.S. (I know, I know — Seyle produced G.A.S. Let's not go there.) The G.A.S., also referred to simply as the stress syndrome, is what he named the process under which the body confronts "stress." According to the G.A.S., the body passes through three universal stages of coping:

- First there is an **alarm reaction**, during which the body prepares itself for the "fight or flight" mechanism. Since no organism can sustain the alarm reaction for too long, a second stage of adaptation ensues.

- In this second stage, a **resistance** to the stress is built. In other words, we can and do become somewhat "inoculated" from stress, to a certain point.

- Lastly, if the duration of the stress is too long, the body eventually enters a stage of **exhaustion**, when it gets worn down by the stressful situation.

One of Seyle's most interesting findings occurred during his work with frogs. He found that if he placed frogs in a vat of water, and then heated this water in tiny increments, that the frogs would continue to sit in it until they boiled to death. This happened even when they had the option of jumping out of the water. Seyle discovered that frogs are extremely adaptable creatures, and can adjust to the ever increasing temperature of the water. However, Seyle also noted that if frogs are initially placed into a vat of hot water, without the chance to adjust, they will immediately leap out of it. What he determined was that the amphibian brain of the frog, while having the ability to adjust to changing conditions, does not have the ability to plan. In other words, the frogs could not set a trigger point as to getting out of the vat when it got too hot. Consequently, they sat in the water until they were cooked.

As humans, we have the ability to plan and to set trigger points for ourselves. We can determine when "the water gets too hot," so to speak. Two things must be in place in order for us to do this.

- First, we must know our own stress reactions. It can be chewing on our fingernails, it can be drinking or eating to excess, it can be a short temper, or maybe it's just becoming quiet and withdrawn. Whatever the reaction(s), if we are aware of them we are in a position to control them. To find out what your stress reactions are, spend some time doing a little self-analysis. Ask yourself what you do when stress builds. If you're not sure, ask people who know you well what kinds of things you do when your stress level builds. If you ask them in good faith, they will probably tell you.

- Next, have a plan in place for what you are going to do when you notice your own particular stress reactions occurring. Maybe a good, healthy plan for you would be to go get a good physical workout to let off some steam. Or, it could be making a list of the things that you need to do in order to get a handle on your overcrowded schedule. Again, every individual is different, so the point of departure is self-analysis. We all probably know what kinds of activities help us to mitigate the stress that we encounter. Then, when we hit our stress trigger point, a plan is already in place to deal with it.

Well, you might be saying to yourself, "That's all fine and

dandy, but what else can I do to help mitigate stress?" Below are some other suggestions on how to handle stress if you are not sure where to begin.

Techniques for Dealing with Stress

If the body is relaxed, the mind will follow. As easy as it might sound, deep and deliberate breathing is a sure-fire way to relax and de-stress yourself. Sport psychologists refer to it as "**Breath Control Training**" or BCT for short. It is a quick and reliable way to manage stress, and can be done in as little as five minutes.

- Initially, begin in a place that is free from any distractions. Sit down, or lie down, whatever is most comfortable to you, then close your eyes.
- Slowly inhale through your nose to a count of four. Try to fill up your diaphragm as you do this.
- Now, exhale through your mouth to a count of seven, feeling your lungs and diaphragm empty out.
- Simply repeat this process over and over for the time you have allotted.
- Focus on inhaling through the nose and exhaling through the mouth.

Inevitably, your mind will wander, and you will find yourself starting to think about other things. If and when this happens, shift the focus back to the breathing. With practice, you will get better at BCT. Then, you can effectively do it even in noisy or distracting environments, like the ones we are used to as firefighters.

Breathing by the Numbers is similar to BCT. It too can be done in very little time. Inhale normally to a count of four, then pause to a count of four. Next, exhale to a count of four, then pause to another count of four. Repeat this same sequence for the necessary time. Again, if your mind wanders, bring the focus back to the breathing and counting.

Progressive Muscle Relaxation is a technique which teaches you how to recognize where specific areas of tension are located in the body, and how to release this tension. (Also referred to as PMR, it is not to be confused with PBR, which stands for Pabst Blue Ribbon, another relaxation strategy used by many.)

When starting PMR, try to find a quiet area that is free from distraction. Lie down on your back. Your feet should be about shoulder-width apart, and your hands placed comfortably by your sides. PMR consists of tightening or contracting one group of muscles at a time, followed by holding that tension for a period of time (ten seconds), then releasing that tension. During both the tightening and releasing actions, concentrate on the feelings in that specific portion of the body.

- Begin PMR by focussing on your legs. Lift them both off the ground just slightly. Tighten all the muscles in the legs, and hold that tension for 10 seconds.

- Release the tension and let both of your legs go limp or loose. Relax for ten seconds while simultaneously breathing slowly and deeply. Repeat the same process one more time.

- Next, concentrate on the pelvic area and the buttocks. Hold this tension for 10, then release. Relax for 10 and concentrate on your breathing. Repeat this sequence again. Repeat this process for the stomach, and then the chest.

- For the arms, make a fist in each hand and focus on

tightening all of the muscles up and down the arm. Release, relax, then repeat.

- With the shoulders, push your back and the shoulders down into the floor and flex them. Hold the tension, and concentrate on the feelings in that specific portion of the body. Release. Inhale and exhale while relaxing, then repeat the process again.

- Now, move your focus to the face. Tighten all the muscles here by clenching your teeth and jaws together. Hold for ten seconds, then release. Repeat the sequence.

- Finally, tighten the entire body for 10. Relax, and repeat. Now, as you lie here comfortably, think about a "relaxation cue" or signal that will help you to easily recall the feelings that you are now having. It can be a short phrase (such as "let it all go") or even just a word. It might be a picture or setting that you visualize (such as a peaceful mountain lake), or even a color.

- Shift your focus to your breathing and the feelings of relaxation in your body.

- Repeat your relaxation cue over and over until it is firmly planted in your mind.

Through practice, this simple relaxation cue will help to bring back the calm feelings in your body. Just like anything, practice makes you more efficient at PMR. Eventually, you can streamline the process down to just a minute or two without having to go through the entire tightening/relaxing process. If you hold stress in a specific portion of your body, focus on that area in particular.

As wildland firefighters, stress is an inescapable part of our jobs. Doing what we do, it is going to be there. It can come from our working or home environments, from our bodies, from our own thoughts, or any combination thereof. However, this does not mean that we have to succumb to its negative effects upon Optimal Performance. If each of us knows what our own stress reactions are, and if we have a plan in place to deal with them should they start to occur, then we are in a much better position to take positive, proactive steps to mitigate these stressors. By utilizing some of the above techniques (as hokey as they might seem, initially) we can more effectively manage stress and become better firefighters in the process.

Transitions

March 11, 2005: I have decided to take the year off from smokejumping. I might not ever jump again or fight fires. For now, I'm taking just one season off, because it's easier to leave a job you love if you are planning to come back. The prospect of leaving home for three months straight to go smokejumping does not square with the realities of my personal life: a wife and an almost-two-year-old boy at home in Minneapolis, 1100 miles away from the smokejumper base in Missoula. By not coming back, I have forfeited my appointment, and the few perks that it provides. My decision is made more stressful by the fact that I have absolutely no idea what I will be doing for work and income. I have fought fires for the last sixteen years, and enjoyed every minute of it, so the thought of missing a season (or quitting altogether) is eating me up inside. Time will tell, I guess. If the 2005 season ends up as busy as it is shaping up to be, I could be an emotional wreck by midsummer as I read the newspaper reports, watch television, and talk with my bros who are still living the dream. What's the deal with unemployment benefits, again?

In the fire vocabulary, the word "transition" can mean a couple of different things. First, it may refer to a specific fire that is growing in size, and thus "transitioning" from one classification to another, for example from a Type IV fire to a larger, more complex Type III incident. Transition can also refer to the process that happens when one management group, say a Type 2 team, turns over control of the fire to a Type 1 team. In both incidences, the term transition refers to a moving from one classification or category to another.

However, the term "transition" can and does have another

meaning within the world of performance psychology. When an individual is in a state of transition, she or he is said to be moving from one phase of their life to another, distinctly different period. As firefighters, as with the fires we fight, we may at times find ourselves in transition. This can happen in different ways.

- Transition might mean transferring from one district to another district. During this change, you might be simply moving laterally; or, in other words, keeping the same job, just in a different place.

- Or, you might be shifting gears altogether. Most of us, at some point in our career, will face the prospect of transitioning from one position to an entirely new and different job title within the fire world.

- Perhaps you are leaving the district engine that you have been on for the past three years and have accepted a job on a rappel outfit.

- Maybe you are vacating your position on the hotshot crew because you have been selected as a rookie smokejumper for the upcoming season.

- Or, perhaps, your transition entails moving from a job as a crew member into a position of fire leadership, be it a squad boss, crew boss, or engine boss/captain. Such a promotion brings with it a change in duties and an increase in responsibilities (not to mention a big increase in pay).

For some people, transitions can be a difficult and stressful time, and so transitions can negatively affect performance if the transitioning person is not prepared for the process. In order to make a successful transition, one must be ready for the changes that come with the new job title. Since the focus of this book is geared toward optimal performance of current wildland firefighters, the process of transitioning from one job to another job within fire suppression is the first concern of this chapter. Later we talk about transitioning out of the firefighting career, be it to retirement or quitting firefighting for another occupation altogether.

Transitions within the world of fire

Through training, experience, and seasoning, firefighters develop a zone of comfort around themselves. They know their co-workers, they understand what is expected of them in their position, and they are familiar with the rules and systems that are in place to help guide their functioning and behavior in the workplace. As mentioned earlier, the biggest challenge faced when accepting a new position is that we are exiting our present comfort zone. This can be a stressful process, since we are trading the devil we know for a devil we don't know.

The key, obviously, is to turn our "unknowns" into "knowns" as quickly as possible when taking a new fire position. Much of this can be done before the first day on the job. When I learned that I had been selected for smokejumper rookie training in 1995, I was both elated and scared shitless at the same time. After five years of trying, and hundreds of hours and dollars spent in the application process, my dream had finally come true. However, I would be leaving behind the security of my crew boss job and most of the friends that I had acquired during six years of working for the Miles City, Montana, District BLM.

Notified at the end of January that I would be in the next rookie class meant that I had four months in which to prepare. Since my overall understanding of smokejumping was quite limited, I sought out as much information as I could find on the subject. I read every book about it that I could find, my favorite soon becoming Norman Maclean's Young Men and Fire. I poured over the training guide that the Missoula base sent to every incoming rookie. I carried a fifty-pound pack three miles into Mann Gulch with my best friend, Jeff Hindoien (see Chapter Three), both to get into better shape and to learn more about this tragic event by seeing the ground on which it occurred. With this preparation came confidence that I needed to make this life-changing transition.

In hindsight, I should have paid much more attention to my conditioning, as I would learn during rookie training that my physical reserves were not on par with most of my other classmates. Too many beer runs, too few long-distance runs. Fortunately, however, I did make it through smokejumper school. By training just enough, and by learning as much as I could about smokejumping, I turned most of the "unknowns" into "knowns", which helped me through the difficult training process.

Because information is power, the key is to finding the information we need so that we can get the power that comes with it. For any new job that you take, utilize as many sources of information as you can to find out what is expected of you. This info can come from a multitude of different sources, including:

- **The position description (PD).** Every job within the federal government has an accompanying position description that goes along with it. The PD will list such things as the major and collateral duties of the position, the knowledge required for the position, supervisory controls (if any), general guidelines for the position, the physical demands of the job, scope and effect of the position, and personal contacts that will be made while carrying out duties, to name a few. If you are taking a firefighter job within state government or the private industry, ask if they have a PD or its equivalent available for review.

- **Human Resources or Personnel Officers.** These folks can be an invaluable asset for finding out other information that is not available on the PD, such as housing availability, required paperwork that is needed, special considerations, and so forth. Their job is to help applicants, so utilize them if you can.

- **Supervisors.** Since you will be reporting to somebody, make contact with your new boss. This person may also provide you with important information, such as training requirements, reporting dates, equipment needs, or other pertinent data.

- **Policy manuals and Standard Operating Procedures (SOP's).** These types of documents shed light on the "system of systems" that is in place for your new job. When you make contact with your new supervisor, ask him or her about how to get a copy of these materials. The more you know about how things operate at your new duty station, the better prepared you will be once you start working there.

- **Firefighter Intelligence Network (FIN).** The world of wildland fire suppression is in many ways a small one, which you know if you have been a firefighter for long. Ask your firefighter friends or colleagues for any information they might have about your new duty station and co-workers. Somewhere, somebody in the firefighter world has some insight that can be of assistance to you, and it's probably only a phone call or an email away. Just remember to take everything you find out with a grain of salt, and be hesitant to form any steadfast opinions until you have actually started work there.

- **Local Chambers of Commerce.** If you are moving to a

new area, the COC can provide information about local realtors, businesses, and attractions in the area. Since moving itself can be extremely stressful, the more you know about where you are moving to, the better you will be equipped to handle the ordeal.

Once you have started in your new position, this learning process needs to continue. Some things to keep in mind are:

- Regardless of what job you hold at your new duty station, you will be looked at as the FNG (Fucking New Guy) by many of your new co-workers. Firefighters by nature can be a rather hard-nosed bunch. Their trust has to be earned, and rightfully so. After all, you could be in a position to make critical life-or-death decisions that might impact these people directly.

Prior to the start of the 2000 fire season, I was helping out with rookie training at the Missoula Smokejumper base. Fred, one of the rooks, told me that he had heard current jumpers form very quick opinions about the incoming rookies. I could not disagree with him, much as I wanted to, because I had been around the base long enough to know he was right. But jumpers are not the only ones who form quick judgments about new hires. From the minute you begin a new job, you are sized up. Have your "A" game ready on that first day, since many might not give you a second chance to change their first impression. And, perhaps most importantly, be in shape physically. Nothing singles you out faster, and more negatively, than being unable to meet the physical demands of the job.

- Confidence is everything. In your new position, you must be confident in your abilities, but not to the point of cockiness. Don't be afraid to demonstrate your skills and talents when opportunities arise, but remember:

nobody likes a know-it-all.

- Learn about your co-workers, their likes and dislikes, their strengths and limitations, their qualifications, and their general fire knowledge and abilities. Find out who the formal leaders are, as these people are above you in the chain of command and it's in your best interest to respect the position they hold. At the same time, discover who the informal leaders are as well. These folks have earned the respect of their co-workers, without necessarily having a formal leadership position. Informal power and authority comes with respect. Find out why these people are respected. As outlined in the teambuilding chapter, cohesion is one of the most important factors in group performance. Since you are now part of a new team, work at integrating yourself into the mix.

- Continue to study and review the policies and procedures that are in place at your new duty station so that you can better navigate the waters in your new organizational climate.

Partial Transition off the Fireline

Due in part to the physical, psychological, or environmental demands placed upon wildland firefighters, many in the ranks transition out of working directly on the fireline into other positions within fire suppression. This is referred to as a partial transition, since while the person is no longer on the line, he or she is still involved in firefighting in some capacity. Truth be told, very few individuals reach retirement age while still "digging in the dirt." Those who do are part of a select group, and deserve our utmost respect due to their vast experiences. But others transition out of direct involvement in firefighting for various reasons, including injury, a desire to spend more time with family and loved ones, or simply a better job opportunity.

Regardless of the reason, such a move involves a transition. The comfort zone and knowns of your old job are left behind, and in their place are unanswered questions and unknowns about your new position. Many of the sources of information that were discussed previously in this chapter are just as applicable here: PD's, HR people, new supervisors and co-workers, policy manuals and SOP's, FIN, and so forth. Utilize them to find out as much as you can about your new job. The same goes for first impressions, confidence, and teambuilding with your new workmates.

On a special note, if an injury has forced you permanently off the fireline, realize that this can be a unique and challenging transition (refer to Chapter 5 for more information). In a study of elite-level college athletes, Kleiber and others (1987) found that athletes who left sport due to injury had significantly lower ratings of current life satisfaction than other athletes. Such athletes feel unfulfilled in their athletic accomplishments, as they did not leave sport on their terms, but rather as a result of injury.

So, if injury should force a firefighter off the line before he or she is ready to leave, this can be a difficult period of adjustment. Many of the suggestions detailed in the Injury, Goal Setting, Stress, and Attitude chapters should be of assistance if such an event has occurred to you. Once again, the world of fire suppression has many moving parts. One does not have to be directly on the fireline to make a positive, valuable contribution to the overall team effort.

Complete Transition Out of Fire

The other type of transition that each of us as firefighters will make is a transition out of the world of fire suppression altogether. For some, this might come in the form of retirement. After twenty years in service (or after the age of fifty), those with appointments in the federal system become eligible for retirement. Retirement is considered a major transition, as that person is moving from one phase of their life into an entirely different period. Transition can also occur when we "hang up our boots." At some point, many will give up their careers as firefighters and move into other non-fire occupations, long before having enough years in service to qualify for retirement benefits. Although we will keep with us those memories of our time spent on the fireline, we will no longer work in the world of fire suppression.

The transition out of the fire world can be a difficult adjustment period. We are, again, moving out of our comfort zones, and into new and uncharted waters. Many of us will refer to our stints as firefighters as "the best job that I ever had." For a variety of reasons — be they family, better opportunity, or retirement from government service — we leave the fire world behind and explore new options. Often believing we might never find a job that we enjoyed more than wildland firefighting, we launch ourselves into a new career or this different phase in life. However, the smell of woodsmoke or the sight of large fires burning on the evening news can bring back a flood of memories and take us back to our time in Nomex.

One of things that I find most fascinating about most firefighters is that each and every one of them seems so multi-talented. It might be carpentry, playing the piano, an advanced degree in business management, or expert skier. You name it. Pick any firefighter, and chances are that person has some skill or set of skills that are remarkable, and yet have nothing to do with fighting fires. The beauty is that these skills can often become a contingency employment plan should the person ever leave the firefighting world. The non-fire skills that firefighters possess will always be there, and sometimes they can be used to support oneself, should the need arise.

As firefighters, we need to continually develop and foster our non-fire skills and abilities. When a person makes an early commitment to a career identity and does not explore other alternatives, this has been identified as "identity foreclosure" by some writers (Marcia, 1978). Many athletes have particular trouble in making life transitions because they only see themselves as athletes, and have not taken the time or put forth the required energy to develop other skills that will sustain them when their competitive playing days are finished. As firefighters, we need to avoid that trap, as well. We need to see ourselves as more than wildland firefighters. By utilizing our other skills, and by keeping in mind that at some point in the future our firefighting days will be over, we will be better prepared to find other employment when transitioning out of fire suppression.

Some other things to consider:

- Firefighters develop a variety of personal skills that are highly valued in the employment world outside of fire suppression. This includes such things as a strong work ethic, the ability to persevere in difficult environments, and working as a member of a diverse team, just to name a few. Reference these functional skills and abilities on your resume, and talk them up when you are looking for a new career or beginning your new career.

- Realize that at some point, you will no longer work in fire suppression. Therefore, take the time while you are still fighting fire to document or collect some of your unique experiences, so that you have something to help you remember your time as a wildland firefighter. These documents or mementos can include photos, video, a journal or diary, patches or pins, certificates of recognition or achievement, newspaper clippings, letters, and so forth. Keep them organized, safe, and accessible.

- Search the web for those sites dealing with wildland fire suppression if you want to keep track of what is going on in the fire world. Check out www.nifc.gov, www.wildlandfire.com, www.firehouse.com, or www.spotfireimages.com.

- To stay involved in the profession, consider joining one or more of the firefighter associations that have been formed. These include the International Association of Wildland Fire (www.iawfonline.org), the Wildland Firefighter Association (www.wffoundation.org), the Federal Wildland Fire Service Association (www.fwfsa.org), the AD Firefighter Association (www.adfirefighter.org), as well as the specialized National Smokejumper Association (www.smokejumpers.com) and the Associated Airtanker Pilots (www.airtanker.com), to name just a few.

In wildland fire, transition occurs when a fire changes in level of complexity or when one type of Incident Management Team hands over control to another. History has shown this kind of transition to be a dynamic and at times dangerous process. In the realm of Optimal Performance, transition occurs when a person moves from one phase of their life into another distinctly different period. This change might be within fire (a new fire job, for example) or outside the profession altogether (such as leaving the fire profession for another non-fire job, or retirement). We have said that these transitions can be stressful and risky as well, because they entail leaving your present comfort zone for an uncertain future. By taking time to plan and prepare, you will reduce stress and better set yourself up for success in your new endeavors, whatever they might be.

PART THREE: ENVIRONMENTAL FACTORS IN OPTIMAL PERFORMANCE

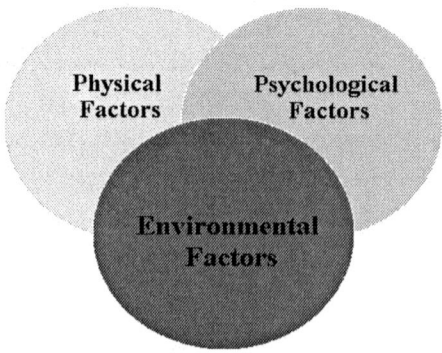

Men wanted for hazardous journey, small wages, bitter cold, long months of complete darkness, constant danger, safe return doubtful. Honour and recognition in case of success

So read the newspaper advertisement, reportedly written by Ernest Shackleton, as he looked for crewmembers for his 1914 voyage to Antarctica, which would eventually turn into one of the most spectacular human rescue dramas in all of recorded history. Despite the grim nature of this "help wanted" ad, over 5000 men applied for the 27 positions available. With a few minor changes, the above notice could very well serve as a call for those interested in jobs within our profession of wildland firefighting:

Individuals wanted for potentially hazardous occupation, small wages, extreme heat, long months of hard work and much time away from home, danger ever present, safe return probable but not guaranteed. Honor and recognition in case of success

What is it that makes such announcements appealing, despite the hazardous warnings, low pay, and general hardship? Just as Ernest Shackleton drew a group of individuals who were willing to look past the danger for the opportunity that the expedition provided, so too does wildland fire seem to appeal to a certain subset of the population. The chance to work outside, with strong and capable people, doing a dangerous and difficult job that is respected by society holds a certain allure for many of us. However, those who choose to stay in the occupation do so knowing that they will work a great deal of overtime, spend a considerable amount of time away from home, and essentially give up their summers (and possibly parts of spring, fall, and winter) because of fire assignments, training, and other work duties.

Wildland firefighting impacts all aspects of our lives, including people close to us. Firefighters often joke that the divorce rate in our profession is 100%. Although to my knowledge no study has been done to check the actual figure for the percentage of marriage failure within our ranks, the fact that we kid each other about it says that the job and a marriage are a tricky combination. Most of us have probably lost a girlfriend or boyfriend at some point in our career, or had strained family relations, due in large part to the schedule demands of firefighting.

As I travel around the country helping to teach leadership courses for firefighters, I am continually amazed at the comments I receive when I ask students about how this job affects their personal lives. Fathers who have missed the births of their children because of fire commitments (although so far, I have yet to meet a firefighter mother who has missed the birth of her child) are not at all uncommon. Birthdays of children and other special events are missed, as are anniversaries, summer weddings, family reunions, and vacations. If it happens during the summer time, chances are good that we will miss it because of the time demands of this occupation.

I am also surprised when I ask students how many of them have friends and family who really do not understand what they actually do as a wildland firefighter. The vast majority of the class will usually raise their hands when asked this question. How, then, I ask, can we expect to get the support and understanding of those close to us when our families don't really comprehend what our job as a firefighter entails?

Talking to firefighters about how this job affects their relationships has revealed an interesting dichotomy. While very few people talk about it during work-related conversations, most everyone appears willing to talk about it once the subject has been brought up in a classroom setting. On most occasions, after I raise the issue of relationships, a group discussion will ensue on that topic. Other times, individuals will approach me after I teach a class to ask about relationship-oriented questions. Many times, they seem to lurk in the back of the room, as if afraid to be seen talking with me about their relationships.

Well, as firefighters we need to talk more about our personal relationships, since our performance on the fireline can be significantly impacted by our functioning on the home front. This chapter aims to do just that. Let's get the conversation started!

11
Relationships

July 22, 1993: The minute the crew gets off the fire and back to the station, I pick up the phone and dial her number. It rings several times before the answering machine kicks in with her recorded message. I have been hearing her message quite often lately, even more so than her real voice, it seems. I leave one more message for her to call me, hang up, and then sit alone in the darkened office trying to figure out what is going on. Something is different, but I just can't put my finger on it. Every time I ask her what's going on, she insists that everything is okay.

Trying to maintain this relationship over the phone is becoming a real challenge. With my girlfriend a thousand miles away, it isn't like I can drop by and take her out to dinner when I get off work. Like a cancer, suspicion is slowly beginning to take over my thoughts. Is she seeing somebody else? Who is she with on these long summer nights when I am stuck here at a ranger station in the middle of nowhere? I try reaching her on the phone several more times tonight, but with no luck. Each time, I quickly put the phone back in its cradle before her machine starts recording.

I only hope things do not continue this way for the rest of the season. We talk, but when we do I can tell she is distant and reserved. Unfortunately, worrying about it is becoming a huge distraction to me. I try not to think about her so often, but it's no use. Inevitably, my mind drifts and I find myself absorbed in thoughts of her.

In September, I turn in my gear and head east, with plans of spending a week with her before starting my other job. When I arrive, I can immediately tell that our relationship is not what it had once been. She tells me that the distance has become too much, and that she sees no future for us. My heart wants to argue that things can change, but my head is telling me that she is probably right. I don't want to give her up, yet I still want to keep fighting fire. These two wishes are like oil and water, I guess.

I gas up the truck and began the long, lonely drive back to Montana. I'll have plenty of time to replay that last scene as the miles of North Dakota pass under my wheels.

The Firefighter Relationship Survey

My relationship with my best girlfriend had begun earlier in the spring, and had progressed in a very comfortable and healthy manner until fire season hit full swing. Face time was supplemented with post cards and phone calls after I was sent out on several details. By fall, though, the relationship was over, leaving us both frustrated — frustrated with each other mostly. We just ended up having two different sets of goals. —"Jose," Helitack member with three seasons of experience

In preparing this book I have interviewed hundreds of firefighters to find out how the demands of their work in firefighting affects the rest of their life. Everybody has a unique situation; some relationships are hindered and some are not. So, in an attempt to gauge just how significant the impact is and what they may have done to mitigate the impact, I have asked hundreds of firefighters to fill out a survey. Throughout the remainder of this section, the actual words of survey respondents will be used in an attempt to convey directly to readers what firefighters go through in dealing with this important part of their life.

The first question posed to firefighters:

- If you are currently in a relationship with another person, or have been in a relationship at some point in the past, how have you managed to maintain this relationship despite the schedule demands of a wildland firefighter (long hours, extended periods of time away on assignment, and so forth)?

Question Two, although similar to Question One in many ways, asked respondents to provide counsel to other firefighters on how to meet this challenge:

- As a firefighter, knowing what you know about the schedule demands of our profession, what advice would you give to other firefighters about trying to maintain their relationships with the people close to them?

The intent of these questions was to find out what kinds of things firefighters have done that seem to work for them in their relationships with others, especially their "romantic" relationships. This is not meant to lessen the importance of their associations with family and friends. In fact, many of the responses also suggest ways to relate to family and friends, generally, in addition to spouses and significant others. The responses are organized into seven major heads:

- **Communication**
- **Partner Qualities**
- **Relationships With Other Firefighters**
- **Time Together/Time Apart**
- **Plans**
- **Priorities**
- **Job Changes**

These seven issue areas form the major headings of this

chapter, followed by a section summarizing what we have learned: **Tending to Relationships — Dos and Don'ts.**

Communication

By far, the most common difficulty pinpointed by the questionnaires is "communication." Of course "communication" is a very general term, so without an agreed definition it is not very useful. Telling someone to "have good communication" is about like telling them to "be safe." Fortunately, many respondents provided detailed examples of what they meant by "communication", and so from their words we can unpack the specifics and learn what they mean.

The most frequent piece of advice from respondents is to buy a cell phone in order to stay in better touch with those close to you. In the old days, talking with loved ones on the phone meant standing in a long line back in fire camp, or looking for a payphone in the little town where you had stopped to buy gas and grab a quick bite to eat. Cell phones have changed all that, assuming we have coverage. Considering how much we are on the road, any cell phone plan you get should include no roaming charges and have as many minutes per month as you can afford.

> *My significant other and I have maintained our relationship through my interest for the job and her family that live in the same geographic location. We have had to communicate much more in the year we've been married. Our cell phone bill has increased significantly.* — "Mitch," 5th year Engine crewmember

Sun Tzu, a Chinese military tactician who wrote the famous work *The Art of War*, taught the principle of Danger/Opportunity, and it is most definitely applicable for cell phones. Although they have many advantages, there is a downside to them. For some of us, ignorance is bliss. In other words, constant communication from the homefront while on assignment can bring with it additional stresses. We almost immediately know about whatever problems those closest to us are facing back home, despite the fact that we are not in a position to provide much help. In addition, regardless of the plan, your cell phone bills are going to be high. Prepaid calling cards that can be used when we have access to a landline can help to mitigate some of this expense.

In spite of the drawbacks, I personally rely on my cell phone perhaps more than anything else to maintain communication with my wife and others I care about when I'm on the road. As technology develops further (better digital cell service, satellite phones, and so forth) it will become easier to stay in touch with those we love. No matter how costly, I believe the price is worth it. In the words of "Vic," a fourth year Engine crewmember: "Cell phones are sweet!"

Another aspect of communication frequently mentioned by firefighters was to educate potential partners about just what our job entails.

> *I have found that my job is a factor that I have to put out there and deal with prior to having a 'serious relationship.' It won't work if you don't tell her early in the relationship.* — "Tom," a fourth season Hotshot

> *Be up front with clear, honest communications. I did not bullshit my wife about intentions or my schedule. Do not downplay your schedule, be brutally honest about the hours and expectations.* — "Curt," an AFMO in his twelfth season

> *You must divulge all info early if you want it to survive!"* — "Julie," 8th season Fuels

When it comes to educating potential partners about what it is that we do, responses seem to indicate that it is a two-stage process. Many firefighters recommend being extremely clear with our spouses and significant others that:

- we travel a great deal
- we can be sent on assignment on very short notice
- we can be out of touch for long periods of time, especially during the summer; and
- we work in high risk environment

Lawyers call this Full Disclosure, and regardless of what we think about their profession, it seems as if our relationships would be well served if we let people who are closest to us know these four facts about our work, whether we are pursuing a new relationship or simply trying to shore up or strengthen an existing relationship.

The second stage of the education process occurs after we have fully and completely briefed our mate or potential mate on the time demands associated with this profession. It applies to our dealings with other friends and family members as well. This part of the communication process entails teaching them about what it is that you actually do. Such an approach has multiple benefits, both to you and those you are close with. Information is power, as we have said before. The more people in your life know about what you do, the more power you give them to understand and plan around your unique scheduling demands and to deal with the everyday realities of your occupation.

> *I take my wife to trainings, when available…and to work when I can. I have brought her and our kids to places that have been important to me, such as old fires that I have been on, and actually to prescribed fires when I have*

gotten permission to do so. — "Rick," member of a Type 2 Team in his 15th season

Surprisingly, perhaps, not everyone agrees that the key to maintaining a successful relationship is communication. An engine crew member writes:

Do not discuss fire. Do not talk about your job. They do not know anything about fire and don't care. They know it is what takes you away from them. — "John," a fifth year engine crew

With all due respect to "John," I disagree. If the individuals we are close with do not care about what it is we do for a living, then I wonder how strong the relationship must be in the first place. Again, when people who care about us do not have adequate information on any matter, they tend to create fantasies to fill the void. Problems regularly occur because the ideas they have to fill the gap are frequently wrong or incomplete. By teaching those dear to us about our jobs as wildland firefighters, we help them get to know us better. Take your family and significant others to your place of work, buy them books on wildland fire, show them the locations of fires you have worked on in the past, or clip out newspaper or magazine articles about your work. Generally, the more people know, the better they can support you.

Partner Qualities

Firefighters often note that certain personal qualities are important in spouses or significant others who must deal with the risks and itinerant character of our occupation. "Mark," a Type 2 crewmember in his third season writes, "Date only strong, independent people who can put up with periods of separation." Mates who can handle a relationship with a firefighter are frequently described as "self-sufficient," "understanding," "strong," "positive attitude," and "supportive." Here's how one veteran firefighter puts it:

> *The person you marry or date must be fairly independent and cannot be needy. If your girlfriend or boyfriend doesn't like the job in the dating phases, multiply that tension by five if you get married.* — "Jack," a fifth year engine crew member

Many of us have known or been involved with "dependent" and "needy" people, or have seen a co-worker trying to deal with such a partner. Obviously it takes enormous energy and time to maintain this kind of relationship considering the demands placed on us by our jobs. Responses to the questionnaire suggest that the best chance of establishing and maintaining a solid relationship is with someone who is independent and secure enough to handle our improvised schedule and changeable work locations.

Another quality highlighted by firefighters themselves is both parties in a relationship must display a high degree of patience. Many respondents caution against rushing too quickly into marriage and having children. "Jack" alluded to this in the previous quote, when he mentioned how marriage can dramatically multiply any pre-existing tensions or concerns. Children, of course, are another topic altogether. Contrary to popular belief, babies do not usually strengthen relationships that are on shaky ground to begin with, but actually add stress and new causes for insecurity or uncertainty.

A key in relationship-building seems to be taking time and letting things develop, be it a new relationship as it moves toward a more serious commitment, or an already developed relationship that is considering marriage and/or children. By being patient in our relationships with others and not rushing into life-changing events (such as marriage and children) both parties involved will have a chance to let things mature, and to get accustomed to the special time demands of wildland firefighting.

Another common theme in the area of "partner qualities" revealed by questionnaire respondents has to do with trust. In his answer to Question One, "Steve" writes,

> *I explain the hazards related to the job and the safety procedures we use to mitigate these hazards, but being away or leaving for an extended period requires a large amount of trust, which can be hard to maintain. I have not found a way to strengthen that trust. That is why I filled this survey out. I hope to help you help me.*

In any relationship, trust is essential if the union is to survive. "Steve" touches upon the importance of trust, which is extremely challenging for us considering our travel schedules. If one or both partners are separated for long periods of time, questions about fidelity can occur. If a couple is not spending much time together, people begin to wonder if their partner is faithful to them while he or she is absent. In the blunt terms of one assistant fire management officer:

> *You have to trust one another and communicate. If you're wondering if she is cheating on you while you're gone, you're screwed - it's already over.* — "Curt," an AFMO

Firefighters need to realize trust issues in any relationship will be compounded because of our travel schedules and changing work locations. So we need to talk specifically with our partners about temptations and our need for trust.

I am amazed how often firefighters tell me they really do not talk much with their significant other about their innermost thoughts and feelings. Again, how can we expect to fully understand our partners (and for our partners to understand us) if we do not communicate these kinds of deepfelt things with them?

Trust is like a paycheck — it has to be earned. If the paycheck bounces, we have a hard time trusting the checkwriter in the future. Specifically, if one person cheats on their partner and gets caught, it becomes nearly impossible for him or her to be completely trusted by their partner again in the future. That is not to say that if there has been infidelity in a relationship that it can't be overcome. It means only that it will be very difficult to restore trust once it is broken. In order for a person to trust us, our actions speak the loudest, not our words. That being said, effective communication is also vital in backing up our behavior.

Since each partner in a relationship often wonders about the other's level of commitment to the relationship, it is important to have continual discussions about your commitment. Tell your partner in honesty that you are being faithful to them while you are on the road. If you have questions about their fidelity, ask them about it. Should you be unfaithful to your partner, have the courage and decency to tell them about it, and then be prepared to face the consequences, whatever they might be. The wildland firefighter community is a small one. You should expect that your partner will hear about your exploits, so it's better that he or she hear it from you.

Relationships with Other Firefighters

I don't date because of my job. I'm hoping to find someone

in fire so that I don't have these kinds of complications.
— "Jennifer," a Rappeller in her third season

These few words from "Jennifer" reveals a conclusion she has drawn about the toll her work has on her personal life. Here is a young woman, early in her fire career, who has given up dating because she wants to avoid "complications" that result from doing what we do. She is actively not looking to be in relationship because she knows how hard it will be to maintain. Her goal of looking for a partner who is also a firefighter is one that is shared by many others. One experienced firefighter echoes her sentiment:

Be in a relationship with another firefighter who can sympathize with you because they have a similar schedule.
— "Eric," a hotshot in his fifth season

Based on survey responses, it appears that many firefighters have opted for relationships with other firefighters, since they truly understand what the job entails. Several experienced firefighters who happen to be women have found this to be true:

I chose another firefighter so they would already understand the time constraints that I face. — "Tammy," also a hotshot in her fifth year

My significant other is also a firefighter, so it makes it easier in regard to accepting and dealing with the demands of the job. — "Mara," a sixth year Engine Boss

I'm lucky that my husband used to be a firefighter (Engines, Type I Crew), so he understands much more than someone who may not have that background. — "Christine," a Fuels AFMO with fourteen years of fire experience

That said, simply being involved with another firefighter is not going to alleviate the many other demands that are placed on a relationship. It may help with some aspects of your life together, but certainly not all. For example:

> *My husband is also a firefighter so that helps with understanding the job expectations, but we still face challenges.* — "Cindy," a Prevention Tech

> *My former husband and I worked on the same Hotshot crew for three seasons. It wasn't until he got out of fire and I stayed in that the problems began. He had a hard time with me being gone for extended periods of time even though he knew the life of a firefighter.* — "Jan," a logistics coordinator with several seasons on both Engines and Hotshot Crews

In the case of "Jan," even though her ex-husband had been a firefighter and knew what the time demands were, the schedule proved detrimental to the relationship when he was no longer fighting fires. Just because one understands something does not guarantee that he or she will be able to deal with it.

If both partners are involved with fire, this can lead to other challenges, as Marty points out:

> *My girlfriend and I have decided to drop everything when I get home to get the needed time together before I have to leave again. If we don't get this time together, then bad things can happen. Things are intensified because she is in fire, thus extending our time apart.* — "Marty," a Type II crew Squad Boss in his fourth season

For those who are in a relationship with another firefighter, and do not work on the same crew or at the same duty station, time away from each other can be magnified, especially during busy fire seasons. It's possible that the two people in a relationship may not see each other for weeks or months at a time. For those who have children, this can become especially difficult:

> *There is an inherent understanding in my marriage because we are both involved in fire to some degree…My primary job is forest dispatcher and my spouse is on a Type 3 Team, so it has proven to be more difficult at times to arrange schedules for childcare. We have to make our priorities clear to each other, such as who makes more money on overtime and who needs to keep up qualifications.* — "Debbie," Forest Dispatcher

In spite of the challenges, many firefighter-couples have succeeded in their relationship. As one long-time dispatcher eloquently puts it:

> *We have managed it because we are both wildland firefighters and are both fully aware of the demands made on us from our jobs. We work to call each other each day via cell phones if possible. We make the most of our R & R days, and in the winter we re-adjust to living together. In short, lots of give and take. For us at least, sometimes distance makes the heart grow fonder.* — "Sue," a Dispatcher in her 15th year of fire

For most of us, our partners are not wildland firefighters. So they do not necessarily possess the intimate knowledge of what our job entails. If we limit our search of potential partners to only those who are or who have been firefighters (as does "Jennifer"), we are excluding many suitable partners. Just because a person has not been a firefighter does not mean that he or she is incapable of loving us while simultaneously accepting what we do for work.

Instead of restricting our pool of potential mates to current or former firefighters, I recommend instead that we look for a person who possesses the already-mentioned qualities of independence, self-sufficiency, and supportiveness. These are the people who will be able to cope with the time demands that are placed upon us. Once we have found a partner with these qualities, our mission will be to educate and keep them informed about what we do, and to convey the love we have for fighting fire. By taking these steps, we set a more solid foundation for our relationship.

Time Together/Time Apart

Sue's earlier comments that "distance makes the heart grow fonder" seems to be an attitude that is shared by many firefighters when they describe their closest relationships. Once again, we are reminded of the Sun Tzu principle of Danger/Opportunity. Instead of focusing on the negatives (long periods of time away, unpredictable schedules, and so forth) we can search for the positives that are presented by our job schedules. One such positive is this concept of absence being used to strengthen the relationship.

> *My wife understood my role as a firefighter before and after we got married, and she supports me 100%. Although it has been hard our time spent away from each other, I believe, has helped to keep our relationship strong.* — "Doug," a Type I team member with 30 years of experience

> *Absence makes the heart grow fonder. After 32 years of marriage, I believe that my wife is somewhat glad to see me go off. She knows that I enjoy it!* — "Wayne," a Fire Management Officer) in his 34th year of fire

> *Our relationship involves understanding of and respect*

for each other's career path. Our time together is quality time that makes our time apart an opportunity to reflect and an occasion to anticipate rejoining forces. — "Jim," a Smokejumper in his twelfth year

It is great for my wife and I because if we spend all of our time together we can tend to butt heads, but with fire we can get little breaks from each other and it seems to keep things fresh. However, I also have a three year old son, a one year old son, and a daughter due next month. This is where it gets hard, to be gone and have the kind of relationship they need and that I need. I'm also about to start on the Hotshots this year, so the true test still lies ahead. — "Carlos," a Helitacker in his third season

The words from "Carlos" show that even with a positive attitude regarding time apart, being away can still be a difficult for our partners, especially if there are children to take care of. Even if you have a partner who is supportive and feels that time apart is good for the relationship, one must caution against complacency. As firefighters, we must continually realize how challenging it is for our partners to be "left behind" when we go out on the road. In many respects, it is harder for our partners than it is for us. Therefore, we need to be as supportive and understanding of their feelings and challenges as we can.

Another way to look at the challenge of our schedules is to focus positively upon the amount of time that we get back during the off season. Granted, this seems to be changing somewhat, considering such things as prescribed fire, training, and the move into more all-risk activities as opposed to just wildfire. Many firefighters just don't have as much time off as we used to. That being said, our schedules usually lighten up once fire season ends. The off-season can be used to solidify and strengthen relationships that may have been strained during the summer months. By deciding that we will take more opportunities to spend time together later, for the time being we may be able to navigate the hardship in the present moment. Once we have a more predictable calendar, it is up to us to spend more of our free time with those we are close with.

Plans

Survey responses indicate that many firefighters feel they have no control over their schedules. This is often the case. Firefighters have to put up with chaos and uncertainty when it comes to schedule. With the ring of a phone, the buzz of a pager, or the wail of the siren, each of us can be sent off on a moment's notice. Other events, like training and prescribed fires, are more predictable in their occurrence, yet they still take us away from those we are close with. When asked what advice he would give to other firefighters about personal relationships, one firefighter bluntly states:

> *Kiss yo summers goodbye. No camping, fishing, softball, and so forth. Seriously, I would tell them that you can't expect to participate in anything with a schedule because you're clueless to your own. You can't really make any long term plans, or plans period.* — "Shawn"

When asked this same question, one veteran firefighter had an equally dim outlook:

> *Stay single. If this is not an option, be prepared that things may be very rough and that maybe this is not the best profession to go into. You must weigh what is important and try to makeup for the unbalances that are caused by this lifestyle.* — "Tory," an Engine crewmember going into his fifth season

Fortunately, it seems that many firefighters are able to deal with scheduling uncertainties by becoming proactive as opposed to reactive:

We try to make a plan for each month of when we will see each other. I try to take vacation time and avoid assignments when we are planning something big. — "Hector," second year Helitacker

Time management is critical for me and my family. I keep a family planner with my wife and update it very often as well as notify her verbally of any changes. — "Keith," Structural firefighter

Before the season started, we sat down and expressed our expectations for the coming summer. We did a monthly review of how we were doing. We planned a vacation in November, which we both shared in the planning of. This was our light at the end of the tunnel. — "Sam," Hotshot Superintendent in his 21st season

I try to maintain an awareness that being away does effect my relationship. My wife and I communicate about this. We try to call each other on the cell phone as much as possible. We make vacation plans on either end of the fire season, which gives us something to look forward to together. — "Greg," an Engine Captain with 12 years' experience

It is really hard for me to make plans. I always think that there is no way that I will be able to stick to them because something will come up. I have recently been trying to work on making plans. More times than not, what I plan is accomplished. — "Henry," Division Group Supervisor on Type II team

Sun Tzu, in his teachings, also advocates for the principle of Can Control/Cannot Control. According to this notion, we must accept things over which we have no power to change. The flip side of this equation is that there are many aspects of our lives where we can dictate the outcome. The truth is that all of us have more control over our schedules than we often admit.

Based on the survey responses, one way firefighters have of wresting control of their time is to make plans with those people we are close with. Be it a vacation before or after fire season, a family planner, or even scheduled time off during fire season, firefighters often have the ability to dictate how, where, and even when they spend their free time. However, taking control of our schedule is difficult if we don't do it systematically. "Henry" (above) notes that planning is the key to getting things accomplished. From his words we can take encouragement to confront the uncertainties of our schedule by engaging in long-term planning as he does.

In order for our plans to be achieved (especially if it involves taking time off during the fire season) firefighters must have the support of management. In the past, many fire managers have not been very receptive to those who sought time off or who wanted to pass on certain assignments due to family or personal commitments. The general attitude has often been that they own you during the fire season — that the job always comes first. But lately, the fire community seems to be getting better at realizing that the wellbeing of its most valuable asset, its employees, needs to be the highest priority.

For the wellbeing of firefighters to be top priority, fire managers and leaders can take several steps. They, too, must plan ahead and be prepared for those occasions when employees need time off. Having a list of qualified individuals in reserve can help to fill open roster spots. Scheduling a few days of down time for the crew during busy fire seasons will give people much needed time off and provide an opportunity for them to make plans with those they are close with. And lastly, the fire community as a whole can continue to improve upon establishing a culture that encourages its members to maintain healthy personal lives.

Priorities

After "communication," by far the most common responses firefighters have to the two survey questions asked revolve around making personal and family relationships a priority in their lives. At times, we seem to fall into the trap of letting our jobs become the most important part of our existence. Through such things as the fire culture, training, and "old school" philosophies, many have been subtly pressured to accept this mindset. Fortunately, this seems to be changing, as is evidenced by many comments received on the survey.

> *Make sure that you take time for your relationships whenever you can, especially during special occasions. Make sure that the people you care about know that you care for them and that your job is not everything to you.* — "Ron," Engine crewmember in his 3rd season

> *I realized that my older mentors did not know their family and kids, and that the job came first before family. They never had a relationship with their family, did not take time with them, became divorced, and did not know their own kids. I swore I would not do this when my son was born. I quit doing large fire and took short term assignments. I also take a summer vacation with the family during fire*

season. I lost my "fire is the only life" attitude and do not miss it. Don't make fire the only thing in your life. If you do, then don't expect a long-term, love-filled relationship with your family, and you also might want the name of a good divorce lawyer. Take vacations, watch your family grow, and remember not to take it (fire) personally. You never started it, so keep your perspective. — "Stan," Fire Overhead Team member in his 29th season

Get a life! There are more important things than going to every fire. How many times do you alone catch a fire/make the difference? — "Donovan," Training Specialist

It's hard. I've been with this girl three years and now she realizes that dating a firefighter isn't as glamorous as it sounded. I have relocated to get closer, and I'm trying to relocate again to get even closer. Something that I just learned is that I love my girlfriend and I like my job. I have reprioritized my life. So now I would tell new people coming in that this is just a job. Put the wife and kids first. — "Brad," a Type II crewmember in his 4th season

In the off season I make myself unavailable for assignments so my family can at least count on me in the off season. I often plan vacations or special events during fire season and commit to them. I'm finding that management's attitude is changing and they usually support taking time off with advance notice during fire season. I encourage people to take time off before the season starts, and spend it with their family. Second, I support my firefighters taking time off in season as long as I have advanced notice of it so I can fill in. Third, don't be afraid to miss a fire. You can still be a committed and good firefighter and not go on every fire. Fourth, management needs to support in season vacations and the message needs to be spread that it's okay and good for you to take time off during the season. Balance between work and family is the key. Who is really going to take care of you in the long run, the Forest Service or your family? — "Lance," Type II crew Superintendent in his 20th season

Family comes first. You don't have to go to every fire. Fires will always be there, but there will only be one first birthday, and so forth. I give leave to personnel on my crew if I'm able to get somebody else or if having my module short won't affect its performance. — "Keith," DFMO in his 27th season

Our families are number one. That should be our priority, and for a lot of years it wasn't mine. But now my boys are grown up. I missed a lot of their life that I will never get back. If I had to do it over again, I would stay more focused on their needs and be there more. There will always be plenty of fire! — "Jeff," Interagency Hotshot Crew Superintendent in his 32nd season

Jeff's words reveal the high cost that each of us can pay if we choose to put our jobs ahead of those people that we care about. Gale Sayers, the great former running back from the Chicago Bears, wrote a book in 1970 entitled, *I am Third*. In it, Sayers writes, "The Lord is first, my friends and family are second, and I am third." Sayers prioritized his life in that way, and put himself after those he considered most important to him: his God and those with whom he was close. Even if you are not a religious person, this advice is still appropriate, because it places you after your significant others.

Job Changes

A hard reality is that if we prioritize our life and discover that our relationships are indeed the most important thing, then each of us might find ourselves in a difficult situation. How can we continue doing our jobs safely and effectively while keeping our relationships as the number-one priority? Unfortunately, this is an extremely difficult question for which there are often no easy answers. Motivations for doing this work vary widely, but one that is broadly held is that we love doing what we do. Since most firefighters are passionate about the job, we may find ourselves in an emotional bind between our passion for the job and our love for the people close to us. For some, maintaining important relationships can mean changing jobs within fire, or leaving the fireline behind altogether. Either of these choices can be very tough.

> *As a firefighter, I'm trying to transfer down closer to home. So try to get a job closer to your home so you can drive home everyday when you are not on fire.* — "Thad," 4th season Engine crewmember

> *Plan on being single unless you find someone who is exceptionally understanding. If the people close to you are a priority, you may want to consider another career.* — "Jim," Fire Use Module leader in his 10th season

> *Get a different job if you don't want to get divorced!* — "Ed," Smokejumper in his 11th year

> *Trust and communication, both characteristics play a vital role in the success of my relationship. I am looking to get out of the profession and spend time with my soon to be wife. My girlfriend has known all along that I am not making a career out of this, it's a means to finish school. Now that I'm finished, it's time to move on.* — "Tony," Engine crewmember in his 5th season

> *First, the money is good and my wife loves shoes, so that*

helps. Second, this is my job and I love my job. Third, I was a firefighter when we were dating so she knew what she was getting into. However, like most women, they like to change you. Fourth, I am not always going to be a firefighter. Once the kids come, I will probably hang it up so that I am around for T-ball and soccer games. — "Brian," Engine crewmember in his 4th season

In fact, being a firefighter has been detrimental to my relationship. I will be starting as a dispatcher next week. I have made this decision as a trade-off because I value my relationship more than fire. Too many people I've worked with have lost their family. You have to prioritize. It is great to be a hotshot, but if you have a family, what about an engine job? Also, consider moving into fuels, up to a battalion, or some other position. At 57, you will retire. What do you want to have? What do you want your life to look like? — "Tammy," starting her third season, moving from engines to dispatch

Changing jobs may be the most difficult choice we will ever face as firefighters. Each of us must weigh the costs and benefits of the profession, and its impact on our relationships, and then decide accordingly. For some people, this means finding a job within the fire community that allows us a more predictable schedule, and thus more time for family and friends. Other firefighters may decide that leaving the profession altogether is best. Or some may choose to keep fighting fire while doing their best to manage and mitigate the effects of the occupation's unique time demands.

The good news is that if you do choose to stay in the profession, you can be sure that many of your fellow firefighters have also managed to make lemonade out of the proverbial basket of lemons. Listen to some veterans tell about it:

I have been married for 29 years. My wife is very supportive of my work and I hers. We have three grown children and one granddaughter. Some assignments have been difficult at times. For the most part, it works out. With the help of God, prayer, and love for each other, it can be done. — "Zach," Engine Captain with 14 seasons experience

If you really love someone, and they love you, you overcome the bad things. It's hard, but you do it. Phone calls just to say "I love you," late nights when you get back, and so forth. You keep each other interested in the best ways you can. A person has to be real and keep open communication with the people you love and care about. — "Alfonso," 3rd year engine crew member

Stay in communication with them on a daily basis. I am a single mom with three sons, 12 years, 9, and 6. I have a great support system with friends and family. When we spend time together, whether it's staying at home or having an activity to do, we hug lots, laugh, smile, tell stories, and talk about what's going on. They do miss me, and I them, but we make it work. — "Julie," Helitacker with 3 seasons experience

I have been married 25 years, and have seen various effects of my job through this time frame. From my wife crying when I left (the early years), to her encouraging me to go for the extra dollars (the later years)! I try to compensate for my time away by purposefully spending more time doing family things when I am home. I also encourage my wife to "get away" and let me stay at home to take care of the kids, and so forth. — "Shane," Type I team member in his 31st season

Remember that leaving is hard, but being left is even

harder. My wife and I promote pride within our family for what I do, so my children are proud of daddy and love it when I come home. My wife is very open-minded and feels that traditionally men went away on the hunt or for sacred rituals. This did not disrupt the family and is natural. You should know that she is also an active feminist and demands equality. It's just that we understand our roles because I am better at providing, and she at nurturing. This is what works for us- to each their own. — "Tom," Engine Boss in his sixth season

As seen from these passages, staying in the firefighting profession can be made to work with having long-term relationships. We can have our cake (our families) and eat it too (our jobs). Having both is not easy, as it requires patience, understanding, communication, and incorporating many of the suggestions that have been provided to us by the firefighters who took the time to fill out this survey.

Tending to Relationships — Dos and Don'ts

Since their comments are so extensive and varied, a comprehensive review of firefighting tips should help to bring their feedback into focus. What follows is a summary of "Dos and Don'ts" for dealing with the impacts (both real and potential) that firefighting has on our relationships.

- Do communicate. Since our work schedules impact our most important relationships, we must talk about it openly with those people we care about. I am amazed how many firefighters tell me this is the most significant issue for them, but that they rarely, if ever, communicate with their partners, family, or friends about it. Part of the answer is that we know it will be a difficult conversation, so we avoid doing it. Well, it's more difficult for us in the long run if we ignore communication than if we tackle it head-on. Ask the important people in your life what their feelings are about your work schedule and the reality that you can be gone for extended periods of time. Find out what kinds of things you can do to make your long absences easier for them. Let them know how you feel about being gone, as well. And lastly, have these types of dialogues frequently. Do not assume that simply talking about it once will suffice.

- Do get a cell phone and a calling card. Expect that your monthly bill will be high, but hey, you can't put a price on love, right?

- Educate the people you are close with about what your job entails. Take them to trainings, give them tours of your workplace, bring them out to old fires, buy them books or videos on wildland fire, clip out newspaper articles, give them a crew T-shirt, and so forth. The more information they have about what you do and where you do it, the better their understanding will be of this profession. Bring them into the mix and make them feel like a member of your fire team.

- Practice "full disclosure" with any potential partner. Let them know the time and scheduling demands of this profession from the very beginning of the relationship.

- When you become interested in starting a relationship with someone, focus on those individuals who are strong-willed, independent, and self-sufficient, and try to avoid getting involved with someone who is overly dependent and needy. The latter type will have a difficult time adapting to your firefighter's schedule.

- Do not rush into anything. Be patient in any new relationship. Try to avoid going too fast, too soon. Both parties need to understand fully what the impacts will be of long periods of separation and unpredictable time demands.

- Trust is the foundation on which relationships are built. If there's no trust, there's no relationship. Talk about trust with your partner, back it up with deeds, and expect that if trust is broken the relationship will be in extreme peril.

- If you are interested in beginning a new relationship, consider dating another firefighter, as they are most familiar with our lifestyle. But realize that this means you both might be gone a great deal and apart from one another. And try not to concentrate your search only on other firefighters, as this eliminates a huge pool of potential partners. Instead, focus on those with the aforementioned qualities of self-sufficiency and independence.

- Try to look at the time apart from one another as a positive, not a negative. Strive for "Absence makes the heart grow fonder," not "Out of sight, out of mind!"

- Become proactive with your schedule. Make plans with the people you care most about, such as a vacation before or after fire season. Take some time off during the fire season to re-connect with those important to you. Maintain a daily/weekly/monthly planner so that your partner knows when you will be gone for trainings, meetings, and so forth. Give your supervisors plenty of advanced notice when you need to be gone.

- Do not make your job the number-one priority in your life if you intend on having healthy and supportive relationships with others. Keep things in perspective and strive for balance.

- When you are gone on assignment, let your loved ones know that you are thinking of them, through such things as postcards, phone calls, emails, text messages, and so forth. Upon your return, bring flowers or other mementoes from your travels. This shows that you have kept the person in your thoughts and taken the time to do something special for them. When it comes to bouquets, "Tony," an engine crewmember in his fourth season, declares: "Bring home good flowers, not Safeway, every time you come home!"

- If you are a supervisor or in a management position, encourage your firefighters to take time off to meet family obligations, be it during the off season or even in the summer months. Avoid sending the message that the job is the most important thing in the world. Be prepared to back fill on your crew when people need time off.

- If you discover that your current position in wildland fire is not providing you enough time to focus on the important relationships in your life, it may become necessary for you to change jobs, either by finding something else within fire that gives you more time at home, or leaving the field altogether. Realize that this may be one of the most difficult decisions you ever have to make.

- Do remember that many firefighters have been able to maintain strong relationships while working in this profession. Understand that it will take dedication, commitment, communication, and a great deal of support from those that you love.

Environmental factors, most notably our relationships with those people we are close with, might very well be the most important, yet frequently overlooked factor of optimal performance. We can become so focused on our physical conditioning and our internal psychological challenges that we forget about the people near and dear to us. By incorporating some, or all, of the above strategies, you will set yourself up for success when it comes to dealing with the time and scheduling demands of the job, and the impact this has on your relationships.

12
Pulling It all Together

Wildland firefighting is a dangerous and demanding occupation. It always has been, and will continue to be as long as humans decide to place themselves in harm's way on the fireground. Since the historic and devastating fires of 1910, more than 900 wildland firefighters have lost their lives in the line of duty. Thousands more have been injured while engaged in or preparing for fire activities. Despite improvements in equipment, training, and safety, these numbers continue to grow year after year.

With these inherent dangers, our jobs require a great deal from us. The body of the firefighter must be strong enough to handle the arduous duties of the profession. The firefighter's mind must cope with the challenges of a high-risk occupation. And lastly, firefighters require support and structure from the people and systems that surround them in their lives.

The fields of exercise physiology and sport psychology have made valuable contributions to the art and science of fire fighting. One particularly helpful contribution is the optimal performance model, upon which this guide is based. Performance includes physical, psychological, and environmental factors, and optimal performance occurs where all three are maximized.

Peak Performance Model

The zone of optimal performance is actually quite small. Optimal performance must be sought and earned, and it does not usually happen on its own. Optimal performance is rare, but is not impossible to attain. In order to safely and effectively complete our missions, wildland firefighting often calls for outstanding performance from those who engage in it. Firefighting often demands that we be functioning in or very near this Optimal Performance Zone.

Just as the fire triangle demonstrates that heat, fuel, and oxygen are necessary for combustion, the Optimal Performance Triangle requires physical, psychological, and environmental factors reach their peak.

Physical factors form the base leg of the model, and are the foundation upon which everything else rests. The bottom line is that wildland firefighting can be and is extremely taxing on the body. Hiking, lifting, digging, sawing, and the other physical activities we engage in all demand a high level of fitness. The Firefighter Performance Workout included in this guide is intended to prepare the entire body for the rigors of the fireline. However, as noted in this guide, physical performance consists of much more than just one's current level of cardiovascular fitness and strength. It also entails proper nutrition and hydration for our internal "engines"; that's why an entire chapter of this guide has been devoted to it. Physical performance also depends on effective treatment and rehabilitation of injuries, should they occur. The book devoted one chapter to this issue as well.

Psychological factors in firefighting are many and diverse. This guide focused on five: Attitude, Goal Setting, Teams, Stress, and Transitions. A positive attitude is necessary in dealing successfully with the many difficult environments in which we work. Goal-setting helps us focus our motivation and efforts. An understanding of how teams develop and what role we can play in the process helps drive group performance. Being familiar with the sources of stress and knowing how to reduce it helps mitigate its harmful effects. And, finally, we presented strategies for those who are confronted with transitions of all kinds.

Confronting environmental factors of performance may be our greatest challenge of all, and experience has shown they may be the least understood and least discussed of the three factors of human performance. While environmental factors, too, is a broad category, this guide focused specifically on the personal relationships that firefighters have with the people closest to them. Due to the time and energy demands associated with the firefighting profession, relationships with family and friends can become stressed and difficult to maintain. To learn more about this, I have asked hundreds of firefighters to complete a survey which focused on the impact of our jobs on our relationships, and what we firefighters do to mitigate these effects. Actual comments from many of the respondents were used to highlight this issue. From their experiences we learn a great deal of wisdom.

If you have read through this entire guide, and are now wondering where to begin in your quest for the Optimal Performance Zone, a self-assessment is a good place to start. Self-assessment can be accomplished by asking yourself a couple questions. Of the three factors that comprise Optimal Performance (physical, psychological, environmental), in which am I the strongest? Your answer to this becomes your anchor point. From this personal stronghold, you will begin your efforts. While this "anchor point" will probably still need some work, it represents the safest and most logical place to start.

The next question: In which of the three areas do I need the most work? This symbolizes the head of the fire, since it is here that you will have to dedicate the most effort and energy. However, there is no point in attacking the head of the fire without first having established a solid "anchor point." If times get tough, you can always retreat back to your anchor point.

Now, so far in your self-assessment, you have identified the performance factor you consider to be your area of strength (anchor point), and the area most in need of attention (the head of the fire). In order to get to the "head of the fire," you will have to work up the flanks. The flanks are simply whatever performance factor you have not yet chosen. They too will need some attention and improvement in order to maximize their effectiveness. Perhaps an example will help flesh this out.

"Jay" considers himself to be in excellent physical condition. He works out several times each week, and stays active year-round doing all sorts of outdoor activities. Jay has always been able to meet the physical demands of firefighting, no matter how difficult the incident or assignment. However, when it comes to his eating habits off the fireline, Jay is a fast-food junkie. On fire assignments, he tends to get by on candy bars and any other junk food he can get his hands on. He knows he should be eating better, but when it comes to nutrition he just does not fully understand how to best fuel himself.

Psychologically, Jay does pretty well in handling the rigors of the fireline. He usually maintains a fairly positive attitude, and for the most part manages the stress that comes with the occupation. He enjoys fighting fire, and he would like to become a career firefighter. Jay has set several goals for himself that, if accomplished, will help him meet this objective. However, Jay at times has difficulty dealing with some of the people he works with. Many have accused him of being too self-centered, and not a "team player." When he transferred over to a new crew last year, the transition was difficult for him. He had a hard time moving out of his comfort zone, and dealing with his new teammates and the "system of systems" in place at his new work location.

From the environmental factors standpoint, Jay faces his greatest challenges. His family does not seem to understand what he does for a living, and why he is so often out of contact with them. Jay has also been dating a woman for the past two years. The relationship has gone pretty well during the winter periods, but once fire season rolls around the tension starts building. She seems to have a difficult time understanding why his job needs to take him away for such long periods of time in the summer. He has tried to talk with her about it, but the conversations usually end up with each of them feeling frustrated. On the fireline, Jay is often distracted thinking about whether or not the relationship will last.

As should be evident from this example, Jay has work to do before he can reach the Optimal Performance Zone. So far he has been able to meet the physical demands faced by a wildland firefighter, but some improvements would lead to much better overall functioning. Although physical performance factors are his anchor point, Jay could make some changes to improve his functioning in this area, such as eating better. From a psychological factors standpoint, Jay also has some strengths. However, his difficulties working in a team environment and dealing with transitions have kept him from maximizing his potential in this area. And, finally, the important relationships in Jay's life represent the area of greatest need. Environmental factors are severely limiting his ability to reach the OPZ.

This guide was designed to provide readers with an understanding of Optimal Performance, and how all three performance factors contribute. This guide is not meant to be the final word on firefighter performance. Rather, its intent is to get dialogue started on a complex and complicated issue, and also to provide suggestions on how to reach the Optimal Performance Zone. The ball is now in the reader's court. It is up to the reader to decide what changes need to be made in life (if any), and what strategies can be adopted to reach the OPZ. I wish you the best of luck in this effort!

Book References

Astle, S. J. (1986). The experience of loss in athletes. *Journal of Sports Medicine and Physical Fitness, 26,* 279-284.

Brown, C. (2005). Injuries: The psychology of recovery and rehab. In S. Murphy (Ed.), *The sport psych handbook* (pp. 215-235). Champaign, IL: Human Kinetics.

Crossman, J., & Jamieson, J. (1985). Differences is perceptions of seriousness and disrupting effects of athletic injury as viewed by athletes and their trainers. *Perceptual and Motor Skills, 61,* 1131-1134.

Davis, M., Eshelman, E. R., & McKay, M. (1995). *The relaxation and stress reduction workbook.* New York: MJF.

Driessen, J. (2002). Crew cohesion, wildland fire transition, and fatalities. *Tech. Rep.* 0251-2809-MTDC. Missoula, MT: U. S. Department of Agriculture, Forest Service, Missoula Technology and Development Center. 14 p.

Festinger, L., Schachter, S., & Black, K. (1950). *Social pressures in informal groups: A study of human factors in housing.* Oxford, England: Harper

Kelly, J., Kaufman, D., & Kelley, K. (2005). Recent trends in use of herbal and other natural products. *Archives of Internal Medicine, 165,* 281-286.

Kleiber, D., Greendorfer, S., Blinde, E., & Samdahl, D. (1987). Quality of exit from university sports and subsequent life satisfaction. *Sociology of Sport Journal, 4,* 28-36.

Kobasa, S., Maddi, S., Puccetti, M., & Zola, M. (1985). Effectiveness of hardiness, exercise, and social support as resources against illness. *Journal of Psychosomatic Research, 29,* 505-533.

Kodeski, K., Ruby, B., Gaskill, S., Brown, B., & Szalda-Petree, A. (2004). Nutritional attitudes of wildland firefighters. *Wildland Firefighter Health and Safety Report*, No. 8, Spring, 7.

Lazarus, R. S. (1984). *Stress appraisal and coping.* New York: Springer Publishing.

Lewis-Griffith, L. (1982). Athletic injuries can be a pain in the head too. *Women's Sports*, 4, 44.

MacCoun, R. J. (1996). Social orientation and military cohesion: A critical review of the evidence. In G. Herek, J. Jobe, & R. Carrey (Eds.), *Out in force: Sexual orientation and the military* (pp. 157-176). Chicago: University of Chicago Press.

Marcia, J. E. (1978). Identity foreclosure: A unique challenge. *Personnel and Guidance Journal*, 56, 558-561.

McDonald, S. A., & Hardy, C. J. (1990). Affective response patterns of the injured athlete: An exploratory analysis. *Sport Psychologist*, 4, 261-274.

Murphy, S. M. (1995). Introduction to sport psychology interventions. In S. M. Murphy (Ed.), *Sport Psychology Interventions* (pp. 1-15). Champaign, IL: Human Kinetics.

Petipas, A., & Danish S. J. (1995). Caring for injured athletes. In S. M. Murphy (Ed.), *Sport Psychology Interventions* (pp. 255-281). Champaign, IL: Human Kinetics.

Puchkoff, J., Curry, L., Swan, J., Sharkey, B., & Ruby, B. (1998). The effects of hydration status and blood glucose on mental performance during extended exercise in heat. *Medicine and Science in Sports and Exercise*, 30(5), Supplement 284.

Rotella, R. J. (1984). Psychological care of the injured athlete. In L. Bunker, R. J. Rotella, & A. S. Reilly (Eds.), *Sports psychology: Psychological considerations in maximizing sport performance* (pp. 273-288). Ithaca, NY: Mouvement.

Rotella, R. J., & Heyman, S. R. (1986). Stress, injury, and the psychological rehabilitation of athletes. In J. M. Williams (Ed.), *Applied sport psychology: Personal growth to peak performance* (pp. 343-364). Palo Alto, CA: Mayfield.

Sharkey, B., Ruby, B., and Cox, C. (August 2002). *MTDC Tech Tip, Feeding the Wildland Firefighter* (0251-2323-MTDC).

Smith, A. M., Scott, S. G., O'Fallon, W., & Young, M. L. (1990). The emotional responses of athletes to injury. *Mayo Clinic Proceedings*, 65, 38-50.

Suinn, R. M. (1967). Psychological reactions to physical disability. *Journal of the Association for Physical and Mental Rehabilitation*, 21, 13-15.

Tuckman, B. (1965). Developmental sequence in small groups. *Psychological Bulletin*, 63, 384-399.

Tuckman, B. & Jensen, M. (1977). Stages of small group development. *Group and Organizational Studies*, 2, 419-427.

Yukelson, D. (1997). Principles of effective team building interventions in sport: A direct services approach at Penn State University. *Journal of Applied Sport Psychology*, 9, 73-96.

Index

Arm training, 67-73

Attitude, 118, 127-135, 154, 163, 186, 200, 206-207, 210, 212, 222, 224

Back/Hamstring Stretch, 28

Back training, 52-56

Borg Rating of Perceived Exertion, 31-33

Caffeine, 99-101

Calf Stretch, 27

Cardiovascular training, 23-35, 37, 77

Chest training, 63-67

Chest Stretch, 27

Cohesion, 8, 146, 147-151, 157-159, 163, 166, 185

Complainers, 130-134

Cool down, 75-76

Core/Strength Training, 24, 36-38

Core Training, 38-45

Ergogenic aids, 98-99

Environmental factors, 9, 10, 15, 112, 221-225, 190

Exercise physiology, 10-17, 112, 221

Firefighter Performance Workout, 6, 23, 35-38, 77-78, 222

Flexibility, 24-29, 35, 76-77

Goals and Goal Setting, 120-121, 124, 136-143, 150, 157, 162, 164-165, 194, 224

Groin Stretch, 27

Group dynamics, 8, 151

Hamstring Stretch, 29

Hydration, 8, 19, 79, 94-98, 107, 222

"I Messages", 132-133

Injury, 11, 13, 17, 19, 22, 25, 26, 53, 108-124, 167, 170, 172, 185, 186

Leg Stretch, 28

Leg training, 46-51

National Interagency Fire Center, 7

Nicotine 101-102

Nutrition, 10, 19, 79, 87-93, 104, 107, 169, 170, 222, 224

Optimal performance, 9, 14-18, 107, 112, 125-128, 159, 178-180, 190, 219-220

Peak Performance Model, 14-16, 221-226

Performance Diet for Wildland Firefighters, 89-92

Performance psychology, 9, 14, 180

Physical factors, 11, 17, 18, 112, 222

Physical training, 7, 8, 18-78, 110, 142

Psychological factors, 8, 112, 125-127, 143, 225

Relationships, 8, 152, 170, 193-220

Rest, Ice, Compression, Elevation (RICE), 112-115

Safety, 7, 89, 103-104, 118, 148-149, 201, 221

Shoulder Stretch, 29

Shoulder training, 56-63

Sport psychology, 9, 12, 112, 123, 147, 221

Strength, 7, 10, 15, 23, 30, 35-38, 46, 52, 57, 74, 77-78, 105, 144, 222, 224

Stress, 8, 91, 95, 123-124, 142, 153, 167-178, 189, 200, 222, 224

Supervisors and Crew Morale, 134-135

Supplements, 91, 99, 102-107

Teams, 13, 14, 82, 116, 126, 128, 141-143, 144-166, 222

Transitions, 126, 179-189, 222

Triceps Stretch, 29

Warmup, 24-25, 77

Work Capacity Test (WCT), 22

To order additional copies of this book, write to:

Birch Grove Publishing
PO Box 131327
Roseville, MN 55113

Email: **sales@birchgrovepublishing.com**
Web: **www.birchgrovepublishing.com**